CPL COX

7530-00-222-3521
FEDERAL SUPPLY SERVICE
(GPO)

ERIC J COX

THE
CHARLOTTE
PRESS

Published By
The Charlotte Press
The Charlotte Press, LLC
229 S. Brevard St, Suite 300
Charlotte, NC 28202

The Charlotte Press books may be purchased for educational, business, or sales promotional use. For information please write: The Charlotte Press; Attn: Special Markets; 229 S. Brevard St, Suite 300; Charlotte, NC 28202; or email Cox3960@Gmail.com

Cover Design by: Eric J Cox
Edited by: Cynthia J Cox

ISBN – 10: 0615319599
ISBN – 13: 978-0-615-31959-9
Library of Congress Number: 2009908489

Printed in the United States of America

For information regarding author interviews and/or appearances, please visit www.EricCox.com or email Cox396@Gmail.com with subject: CPL COX.

To Mom & Dad

To the Troops and their Families

ONE

Chapter One

"DON'T TELL THEM I'M A MARINE! I MEAN IT, MOM!" I remember shouting at her as we waited for a table at an oceanfront restaurant in Wrightsville Beach, NC. It was June 4th, 2003—less than 24 hours after my return home.

Nearly five years later in silence and solitude, I stood in front of the mirror at the position of attention in my Marine Corps Dress Blues. I had just finished mounting my medals and ribbons on my chest. I thought back to that day in 2003.

"What was I thinking? Mom couldn't have been more proud of her son, a U. S. Marine, who had just returned home from war. Yet I yelled at her in front of our family? What was I feeling to make me act so out of character?"

A few weeks ago, I was trying to decide what to wear to our 2007 company Christmas party. I work for myself in a very professional yet fun real estate office with great people, and I wanted to dress up for the occasion.

"I know, I'll wear my Dress Blues," I thought to myself. It had been over five years since I had worn them. Excited, I called my girlfriend to tell her my idea.

"Don't do that, Eric! People will laugh at you," she replied.

"Are you kidding me?" I asked. "The Marine Corps Dress Blues are one of the sharpest and most respected uniforms in the world!"

"I know that, but everyone knows how you feel about the Marines. They'll think you're mocking the company!" she answered.

"Ouch."

I guess I hadn't had the fondest memories of the Marine Corps, but honestly I couldn't tell you why. I actually have quite a story and should be proud to tell it. But who would want to listen?

I didn't have an answer to this question at the time, but my thoughts were leading me in a new direction—a direction that felt a little odd but was filling me with a strange new sense of urgency. Whether I wore them to the Christmas party or not, I had to get my Dress Blues out and decorate them.

As soon as I got home that day, I went to my closet to find them. They were still hanging on the same hanger and in the same plastic cover they had been in when I picked them up from the cleaners five years ago. I uncovered them and ironed out the crease that had formed in the trousers from hanging at rest all those years. The anodized brass buttons and Marine Corps emblems had a light film of dust over them, so I wiped them clean with a soft handkerchief.

"Wow," I said to myself as I looked the uniform over. "Now I've got to find my belts, medals and ribbons!"

I had recently moved into a new house. It was close to my office, and was a good investment. I got a really good deal on it because it was a "fixer upper," and since work was being done, most of my things had been put in storage out in the shed.

It had already grown dark outside, so I took a flashlight and went to see what I could find. It was difficult trying to hold the flashlight with one hand while trying to plunder through the boxes and packing materials to find my accessories with the other. I decided I was making more of a mess than anything, and that I would just have to wait until the next day when I could see better in the daylight.

As I was turning to leave, I spotted a cardboard box with a picture of an American flag on it. I looked inside, and found only a certificate of authenticity stating the flag the box had once contained had been flown over the nation's Capitol. Abby had ordered and given me an American flag after I had returned home; where was the flag now? I decided to stay out in the shed and look for it. It was a much bigger item and even in the dim light should be easier to find. In a box of clothes I had long ago put away, I found it. It was folded up in a rectangle fashion, and because it had been stored in the box of old clothes, had been wrinkled severely. I was happy I had found it, but disappointed it had been treated this way.

I took it inside, unfolded it and dusted it off. I dabbed a hand towel in some water and cleaned all the spots of dirt. Next was getting the wrinkles out, so I placed a towel over it and began ironing. Now it was ready to be cased, so I folded it proper, into a triangle.

I sat it on the nightstand and remembered I had stored some boxes in there, too. I opened the drawer and began searching for the accessories to my uniform. I found one of my Rifle Expert badges, but it didn't go on the Dress Blues Alpha uniform—only medals and ribbons. So I placed this badge on top of the nightstand along with the American flag.

The next day, I got up but didn't go to the office. Instead, I went outside to my shed to continue my search. Diezel, my three-year-old Rottweiler, couldn't have been more excited that I was staying home with him.

I thought I remembered putting all my Marine things together in a large white plastic bag. As I began digging through the boxes and containers, I very quickly stumbled upon a box that Abby had given me. It was a box she used after I returned home from Iraq to organize all the letters I received while I was away at war, and even some that trickled in after I had returned. The letters were all there in their original envelopes, organized in chronological order by the postmark date.

Next to the box was a green book the military uses to keep information like logs, records, and inventories in. These books aren't made very well, but they're free from a friend in Supply. They have a hard cover and are wrapped in green fabric, so they are tactical. Marines like to use the books for information such as platoon rosters, chains of command, agendas, periods of instruction, etc. They also like to doodle all over them when there's nothing to do. Plus, if a Marine has one in front of him, it makes him look busy even if he's not.

Underneath this green log book was a diary that Abby had given me before I left for war. She had intended for me to write in the diary while I was away so I would not forget bits and pieces of things that happened and could return home to tell her the whole story. She apparently felt my letters would shelter her from the truth, and I would write only about positive things and not of fear or danger.

I then came across the white bag I was looking for. It had been knotted at the top. Untying it and opening it, I found my khaki belt for my trousers with the bright shining anodized brass buckle still intact. I found a small zippered silver bag that contained the white belt for my jacket and the large anodized brass eagle, globe and anchor belt buckle.

Also inside the large bag was another small white plastic bag from the Marine Corps Exchange. In it were all the medals and ribbons I had been awarded. They were still in their own individual packages, unopened.

"Why had I never opened any of these?" I asked myself.

One of the medals was a Global War on Terrorism Expeditionary Medal that was awarded to us after Operation Iraqi Freedom. There was a Combat Action Ribbon awarded to us for our service under fire in An Nasiriyah. Another was a Presidential Unit Citation awarded to our unit by the President—an award not presented since Vietnam.

I gathered up these items and took them inside. I mounted the medals and ribbons on my jacket and tried on the trousers. They fit a bit tight but only I could tell that—they still fit. I put on a white undershirt and then put on my jacket. So far so good; now I had to button up. I was worried about the waist and the neck, but although it was a little tighter than I would have preferred, the jacket buttoned right up.

I was so excited, but I didn't know why. I thought for a minute I was excited because the uniform still fit me. But then I realized I was excited because I was proud. I had always been proud to be one of the few, a Marine, hadn't I?

I couldn't help but look over at my diary, the green log book with "CPL COX" written in bubble letters across the top of the front cover, and the box full of letters. The books and the box had been closed since June of 2003. The story written on the pages had never been shared.

I wasn't proud of what I had done, seen or gone through as a Marine, but I wasn't ashamed, either. So why had I refused to tell the story? I honestly didn't know. I wondered … by reading my own words, would I be able to understand my own answer? I opened the log book.

The first entry was dated Feb. 24, 2003. It was an explanation of why I had started using the log book rather than the diary so Abby would understand.

"So I have begun in this book as my new journal for fear of losing my other diary if we undergo chemical warfare. For the time being, this will only be used as a notebook, to take any important notes given or received."

Enclosed in the book were several papers. One was a hand-drawn map of a mixed-use land development I wanted to some day build and own. There were medical evacuation (MED EVAC) procedures, a platoon roster, and information relating to daily challenges and passwords for Camp Matilda in Kuwait. There were color definitions of warning flags and flares. There was a Hometown News Release that we were supposed to fill out and turn in so our unit could send it back to our hometowns for publication in the local newspapers. There was a vehicle identification number and personnel inventory for the 7-ton truck I rode in through Iraq. There were notes from an Arabic 101 class that included basic words and phrases and their Arabic phonetic translations. And there were notes from a Rules of Engagement class that included, *"General Guidance: Don't shoot anyone who doesn't need to be shot."*

The remainder of the book was used as it was intended—a log, although this wasn't an ordinary log. It was a personal journal with specific dates, times, names, places, thoughts, feelings, and actions. It was everything I had purposely forgotten.

The story began.

Chapter Two

February 6, 2003: Scattered rose petals formed a trail—affectionate yellow and passionate red—leading from the doorway to the bed. The soft romantic glow of the glimmering candles warmed the room. A vase of beautiful fresh cut roses sat on the table—nine red and three yellow.

Abby was standing there waiting for me. She snapped a picture of the astonishment on my face. It was to be our last night together—our Valentine's Day—although the real Valentine's Day was actually eight days away. She gave me a white gold and diamond ring to wear while we were apart to help keep the thought of her close to my heart. It was one of the sweetest surprises I had ever received.

Although Abby had come into my life just three and a half months earlier, I believed we shared the kind of love that existed only in fairytales. We had met at my friend Patrick's apartment after a night out on the town in Uptown Charlotte. Aaron, one of my best friends from high school, was also there.

At the time, Aaron was working as a bartender at the Aqua Lounge where I was a regular. We had known each other since middle school, when he first moved to my hometown of Asheboro, NC, from Ohio. We graduated with the Class of '99 but after that, our paths separated. He had been accepted at the University of North Carolina at Charlotte (UNCC); I was off to Marine Corps boot camp at Parris Island, S.C. I then spent a year in California for further military training, but when my intended career path did not turn out as I had planned, my MOS (Military Occupational Specialty) became a "2141"—an Amphibious Assault Vehicle (AAV) Repairman. For that particular MOS, there were only three possible permanent duty stations—North Carolina's Camp Lejeune, Camp Pendleton in California, or Okinawa, Japan. I was fortunate to be one of only two in our class who were granted their choice of stations; my new friend, Russ, was the other. In February 2001, I moved back to North Carolina, and was pleased to learn that Russ would be joining me.

When I began renewing old friendships, I learned many of the people I had known prior to joining the Marines were living in Charlotte—a 4-hour drive from Camp Lejeune. I began visiting whenever I could get away. I found things had changed between Aaron and me because of our circumstances, and we weren't as close as we had been through high school. He had joined the Kappa Alpha Fraternity and now had new friends. Because of our lifestyle differences, at first I often felt I did not fit in. But with each visit to Charlotte, I became more at ease as he introduced me to an ever-widening circle of friends. That was when I met Patrick. Patrick is very laid back and easy going, and great to be around. Plus, his roommate Amy had posed for Playboy, giving a lot of us guys even more reason to find any excuse to visit. In the short time I had known him we had become great friends.

After the tragedy of September 11, 2001, as a Marine, it seemed that I began to gain a deeper respect from the people I met. I was made to feel welcome. I was beginning to run with the "In Crowd." I didn't have to wait in lines to get into bars and clubs. I didn't have to pay a cover charge, and I seldom paid for drinks. I was invited into the VIP areas of several exclusive Charlotte clubs.

By the middle of 2002, as Iraqi war rumors began to surface at Camp Lejeune, I was visiting Charlotte every chance I got. Often Russ and another member of our unit, Ryan, would go with me. We stayed at Patrick's place during our visits, but his apartment was mainly a place to keep our stuff. We were always on the go and living in the fast lane. Friday evening, after my Marine unit was dismissed for weekend liberty, I would drive to Charlotte and get there in time to go out that night. When most clubs closed at 2:00 AM, we would go to "after hours" clubs that stayed open until daylight. And when those clubs closed, we would go to someone's house or apartment for another after-party. Sleep was over-rated. We would usually nap Saturday afternoon, but were back at it Saturday night to do it all over again.

It was as if I was living in two separate and very different worlds. Sunday night I would leave Charlotte between midnight and 1:00 AM to drive back to Camp Lejeune and arrive just in time to make it for 5:00 AM Monday reveille. I took ephedrine and caffeine on the way both to stay awake for the drive and have the energy after reveille to make the routine five to seven mile run with my platoon. After work on Monday, I would skip evening chow and go straight to bed, sleeping until the next morning in order to catch up on the sleep I had lost over the weekend.

In the fall of 2002, as the Bush administration continued to build their case for WMD and "regime change" in Iraq, our Marine units were instructed to ready ourselves for overseas deployment at a moment's notice. We began receiving actual dates on which we were supposed to leave, but those dates continued to come and pass without action. On several occasions, when we had been told we would be leaving on a certain date, my new Charlotte friends bid me farewell, only to learn the following weekend that our departure had once again been delayed. Although I felt I was probably losing credibility with them, I kept going back every weekend to party it up before having to deploy.

With all the talk of war, I didn't know when or even if I would be coming home. It was a difficult period, and I was determined to make the most of whatever time I had left. And then, to make the thought of leaving even tougher for me, I met Abby.

It was Friday, October 25, 2002. We were introduced at Patrick's place that night after hours. She didn't have much to say to me. She was arguing a lot with Aaron; Aaron loves to argue even if he knows he's wrong. I have no idea what their arguments were about, but I do recall exactly what Abby was wearing. I remember because she asked me to tie her red shirt in the back to make it tighter in the front. I tried to act as if I had tied a million shirts like hers and it came to me second nature, but I felt strangely nervous. To tell the truth, I had never tied a shirt for someone and I was usually the flirt that *untied* the shirt and laughed about it. I tried to do it just right, but instead tied it way too tight. She then pointed this out to everyone and joked that she could not breathe, making me blush with embarrassment. I couldn't believe she had made me react that way.

At the time Abby and I first met, I was happy and content being unattached and had actually grown quite fond of it for the first time in my life. Eight months earlier, a four-year relationship with my high school sweetheart, Tammy, had ended. She broke my heart on Valentine's Day that year (another reason I was enjoying visiting Charlotte so much; my Charlotte friends played a huge part in my getting over the devastation). And in addition to not wanting to begin dating again, I knew it was only a matter of weeks before I would be deployed overseas.

But when you have the feelings I began to feel for Abby, it's much easier said than done to just walk away because the timing isn't as perfect as you'd like for it to be. I couldn't stop myself from calling her during the week, and I would always get so excited after talking to her. I was getting to know and like her more and more with every conversation. My reason for visiting Charlotte quickly changed from joining friends and partying to being able to see and spend time with Abby.

Abby was also a student at UNCC, majoring in Theater and hoping to become an actress. To help get through school, she was bartending at an exclusive high-energy dance club called Mythos. I had been there several times, but spent most of my time dancing to house music in a big room there called The Oracle. I wondered why I had never seen Abby there, but then learned she tended the VIP bar located downstairs in the basement. I had been to that bar a couple of times with Aaron, but they didn't know me unless I was with him. I didn't want Abby to know I couldn't get into the VIP area without Aaron, so I managed to become acquainted with the bouncer and he allowed me in.

Now on the weekends, instead of going to see Aaron and the staff at Aqua, I was going to see Abby at Mythos and meeting the Mythos staff. I would stay and wait for her to get off work shortly after 4:00 AM so we could hang out together at someone's party. It was only a matter of time until I was back in a relationship again. But I began to notice something. Each weekend when I drove back to Charlotte to see her, it was like I had to win her over yet again. I was falling head over heels in love, but for some reason, she seemed to be holding back her emotions. It felt as if she was afraid of losing control and falling in love with me, and I finally found the courage to ask her about it.

"What are you afraid of?" I asked.

"What do you mean, what am I afraid of?" she mocked. "What kind of question is that?"

"I'm sorry ... please don't be offended," I replied. "It just seems to me that you're afraid of something and it's causing you to hold back."

This short dialogue quickly developed into a long conversation that unlocked the door to our relationship. She was afraid of being hurt, and didn't want to believe any good man existed in this world. She even told me she had once written down everything she hoped to find in that special someone and I was everything she had dreamed of, yet she refused to allow herself to believe it.

The breakthrough proved to be a turning point, and our relationship began to build momentum. We were up late talking on the phone every night when I was away at Camp Lejeune; we were in each other's arms when I was in Charlotte. On November 22, 2002, we announced to our friends that our status was officially that of "boyfriend and girlfriend." Abby went home with me to meet my entire family over Thanksgiving. Shortly thereafter, I met her mother and father in Washington, DC, where her father's business was holding their year-end awards gala at the Ritz Carlton. For me, it was an especially cool occasion ... not only did I get to meet her parents, I also had a chance to wear my Marine Dress Blues! As every Marine knows, that uniform draws attention anytime and anywhere it might be seen.

We both knew that because of family expectations, Christmas was going to be a challenge. In my entire life, the only Christmas I had missed with my family was the year I had been in California and was not allowed leave. And in view of the uncertain future, it seemed particularly important this year for me to be at home. Her family had similar traditions. So as much as we both hated it, we decided that she would travel to Indiana (along with Zula, her beloved Yorkie), and I would stay here in North Carolina for the holiday. I really wanted to send her some type of surprise to let her know how much I missed her, so I called her parents a couple of weeks in advance to let them know what I was planning and get their address. Then I called a local Indiana florist and arranged for roses to be delivered to her on Christmas Eve. She sent me a picture of her holding the roses with tears running down her cheeks. When she returned and we were reunited, all was well again. We rang in the New Year together with friends at the Aqua Lounge. Aaron was serving up the drinks. At the time, it seemed like the world was in perfect harmony.

And then the word came down. The final decision had been made. Hundreds and hundreds of AAVs and support vehicles were shipped out, seemingly overnight. We were advised to get our personal affairs in order, and to be sure that important documents such as wills and life insurance beneficiary forms were updated and accessible should the need arise. War was apparently imminent. Troop deployment for our unit was scheduled for early morning on Friday, February 7, 2003.

Tomorrow is the day I have dreaded for well over a year now. Abby and I knew it would eventually come, but had tried our best to put it out of our minds so we could enjoy the time we had together. Now, it was different. It was for real. We made love, but the unknown made sex and conversation seem distant. We only wanted to hold each other close—closer than ever before. We didn't want to fall asleep; we knew the next thing we heard was going to be the alarm signaling our physical separation and my departure. At this point, it was only a matter of hours.

Chapter Three

At some point during the early morning hours on Friday, Abby and I had both apparently drifted off to sleep. We awoke with the startling realization that we had overslept. We grabbed all of our stuff, ran out of our room, left the motel without even checking out, and raced to the base. A Marine doesn't care about speed limits and traffic laws when running the risk of being late for a formation. In fact, with the importance of the morning, I was actually hoping I would get a ticket so I would have a legitimate excuse for being late. It seems strange that when I want something bad to happen it rarely ever happens, but when I'm least expecting it, I see the blue lights in my rear view mirror.

But even though I was late, it wasn't a big deal. For once, they seemed to understand it was difficult even for a Marine to meet precise schedules when so many of us had friends and family there to see us off and show their affection.

The weather was cold, damp and dreary, and somehow seemed appropriate for the task at hand. Few words were being exchanged, even with all the people there. It was a situation where no one seemed to know what to say or how to say it.

I had thought about this day for months and months and what I would say to those I loved if it ever did come, but when the day finally came, the words just didn't feel right.

I spent most of my time wondering what I could or should be saying. I contradicted my thoughts with other thoughts. I wasted precious time arguing with myself in my mind about what to say, knowing the frustration I would feel when my loved ones were left behind and I didn't believe I spoke the right words or feared the ones I did speak could have been misinterpreted.

But honestly, on a day this difficult, it may be this works exactly as it should. I believe the tender looks, silent moments and warm embraces between loved ones can sometimes communicate feelings that words might not accurately convey.

Blank stares seemed to represent the fear of the unknown that everyone was feeling. It crossed my mind I might never see my loved ones again. I tried not to think it or give any indication I was thinking it, but it did cross my mind. I was sure it was in their thoughts, too. It was a really scary thought; it made conversation and laughter forced and uncomfortable.

I told myself I had to be strong and show no weakness; if any weakness were to seep through to the surface Abby and my family were certain to sense it and worry even more. So I walked with a smile, and laughed and joked with my friends and told people about the guns I was carrying. I told them funny stories about the Marines I was deploying with in an attempt to lighten the mood. It helped for a little while, but then seemed to make the situation even more painful.

On top of it all, as if my gear wasn't hard enough to manage, I needed to keep up with the people that were there to see me off. I had to make sure I was not leaving anyone out. Mom, Dad, my brother, David, and Abby were all so sad to see me leave, and I felt I needed to entertain all of them equally and accordingly.

Mom was a wreck, just as I expected. Mom is my #1 fan in this world, but she never feels she can do enough for me and sometimes it can be overwhelming. This morning she opened the trunk of her Lexus to reveal everything from candy to inexpensive wristwatches she thought might be of some use to us along the way. There was enough to go around to all my friends, so they loved it. All I really wanted were her words of encouragement telling me everything was going to work out in the end and, more importantly, that she was going to be O.K. while I was gone.

I have put my Mom through a lot in my life, and joining the Marine Corps was one thing she asked me not to do. I told her it would be good for me. I assured her she had nothing to worry about, and reminded her she survived through the days of my motocross racing, snowboarding, and jet skiing. Then there was the late night partying in high school, and wrecking my first car ... a car that Dad had spent months and months restoring. Before that there was BMX racing, skateboarding, roller blading and roller skating. All the cuts, nicks and scrapes as a kid and the broken bones, dislocations, and concussions as a teenager.

Now there I was, standing in my desert cammies hugging her with an M16 Assault Rifle strapped across my back, the bayonet affixed to my belt, my gas mask around my thigh, and my M240G Machine Gun by my side. Honestly, I'm not sure how I have managed to put her through what I have and truthfully, I didn't know if she was going to be able to handle what I was about to put her through. She had every reason in the world to be a wreck right now, and there was nothing I could do or say to make her feel any better.

Dad, on the other hand, is a different story. What I wanted all my life was his approval. He has to be the poster child for "perfect," and I didn't think I ever measured up to his standards. He has his B.S. in Mechanical Engineering from North Carolina State University, and has always had a very respectable job that enabled our family to live well and have nice things to be thankful for.

But I always hated schoolwork. During my middle and high school years, there was a lot of disciplinary action taken because of my schoolwork—or lack thereof. I rarely did homework if it took more than 10 or 15 minutes of my time. I didn't see any reason for me to read much of anything other than magazines related to my outside interests. I had no use in my life for history; history dealt with the past and I live my life for today and look forward to tomorrow. Science was fun when we were able to do experiments, but that was rare. And when we did do experiments they made us wear all kinds of protective clothing and that took the fun out of it. Math was fun, too, because it came very easily to me and required no additional studying on my part. I could do my Math homework at the beginning of class before the teacher could take it up.

What I did like in school was having a good time and doing things other than schoolwork in class. This of course got me in trouble on more than one occasion. I didn't really mind getting in trouble; it was actually quite fun because it made me the center of attention—that was, until they began making me write my own letters home to my parents to have them signed and returned. Then it began to get tough because if I didn't get Mom or Dad to sign the letters, both the teacher and I or the Principal and I had to call my parents together and I would have to explain what I had done. This puts a lot of stress and pressure on a young kid that really wants his parents' approval. So I acted as good as I knew how but I still got in trouble ... it's just who I am.

Well, this wasn't who Dad was, or is, and I knew it. Dad is a firm believer in the traditional way of thinking—you go to school and get good grades so you can get accepted into a reputable college so you can get your degree, land a good job and work your way up the corporate ladder. This was what he had done when he began his career at the General Electric plant in Asheboro, which was later bought out by Black & Decker. So for me to be failing at the first part was not in my best interest. I must admit my parents seemed to recognize my nature was somewhat free spirited and didn't punish me as much as they might have, but I can't think of a punishment worse than seeing the look of disappointment on their faces when I had gotten in trouble.

Later on in life, Dad and I grew closer as both our lives were changing. I elected not to go to college with all my friends, and decided the Marine Corps would be good for me. My Mom's thoughts aside, Dad supported me once I had made my decision. He warned me, "Boot camp is going to be the hardest three months in your life. You're going to feel like you're in Hell." He had served in the Army back in the Vietnam era, so I knew he was speaking the truth. At the same time, this was very encouraging to me because I knew if I could do it and succeed, I could win his approval.

In 1999, during my last year of high school, my Dad experienced some very unfortunate events in his life—events which made him a real person in my mind. In June, he lost his job at the plant where he had worked for twenty-six years because Black & Decker sold their Housewares Division and operations were moved to Mexico and China. Dad wasn't old enough or financially in a position to retire, and in his mid-fifties found himself unemployed and looking for a new job. The pension benefits he and Mom had believed they would receive upon his retirement were reduced by two-thirds. His traditional way of thinking was becoming a thing of the past.

At about this same time, I elected to join the Marine Corps. I had taken up the sport of motocross at the age of 14; in my senior year of high school I was racing competitively and at the point that with a little luck and lots of hard work, I could have possibly moved into the professional rankings. But I was going to lose health insurance coverage under Dad's plan at the age of 19, just as Dad was losing his job and income. I knew my hopes for a career in motocross had to be put on hold.

And then one day in October 1999, when I was away at boot camp and basically "out of touch" with the rest of the world, my Dad decided to take my two Yamaha race bikes up to Reidsville, NC, where my sponsor, David Howard, had his shop. Dad was driving "Baby"—my dark blue 1996 Chevrolet Silverado 4x4 truck which I had customized with a suspension lift kit, Flowmaster exhaust, mud terrain tires, sound system, chrome step bars, etc. My life prior to joining the Marines had been centered around that truck and my bikes.

As Dad was driving up the interstate, another driver going in the opposite direction with his newlywed wife and his sister-in law fell asleep at the wheel and ran off the road. Startled, he woke and tried to pull back onto the road but over-corrected, sending his car into a spin and across the median. My Dad was approaching at that very instant, pulling a box trailer complete with cabinetry, spare parts, 65-gallon water tank, and my two bikes. Stopping the full-sized Chevrolet with all the bells and whistles already required some additional effort; needless to say, stopping the truck while towing the trailer required more than just additional effort—it required additional planning.

The out-of-control Nissan Maxima collided with my Dad at a combined velocity of about 100 miles per hour. As one might imagine, the momentum was working in my Dad's favor so he luckily walked away with only bruises. The driver of the Maxima and his wife went to the Emergency Room in critical condition and unfortunately the sister-in-law in the back seat was killed.

My truck was towed to the body shop to have estimates done to determine whether or not it was totaled. Surprisingly, the trailer with my bikes inside sustained little damage but was towed to an impound lot until it could be retrieved. That same night, thieves broke into the lot and got into my locked trailer. They apparently jumped on my bikes, did donuts in the gravel behind the trailer, and rode off. My bikes were never recovered.

Meanwhile, I am in boot camp at Parris Island with no access to the outside world other than by pen and paper. This was also everyone else's only way of contacting me. But as luck would have it, the next day my platoon won first place in a company drill competition and as a reward, we were each allowed a 2-minute phone call home.

It was about suppertime. Since I knew my parents, not expecting a call, would probably be eating out, I surely didn't want to waste my one call on a "no answer." So I tried calling Tammy, my girlfriend, and got her. The timer was ticking, so I had to talk fast and tell her I was doing fine and that I missed her, so forth and so on. With about 10 seconds left in the conversation she quickly added, "Oh, by the way, your Dad wrecked your truck and your bikes were stolen ... but he's O.K." At first I thought she was joking, but it didn't make sense for her to joke on a 2-minute call. Just as I realized she was serious, my Senior Drill Instructor tapped his watch and said, "Time." So there I was with hundreds of questions but no time to ask even one. All I could tell her was that I would see her soon and to tell everyone "Hello" for me.

I immediately transferred all those questions to paper and as I was writing, it hit me. "My Dad is a real person!" So many times in my life I had something happen to me that was completely out of my control. But when I tried to explain myself to him I would feel as if I was being questioned for some sort of confession. The thought of Dad having something happen to him that was out of his control overjoyed me! My truck was probably totaled and my bikes were stolen, but I couldn't have been happier.

Strangely, because of this awful accident, my Dad and I became closer. Now I see him as a real person and not the perfect man I saw as a child. Now I realize he is perfect because he is not. Things have never been better between us. And I proved to him and to myself that I could achieve anything I set my mind to ... I became a Marine!

This morning I was reminded my Dad is not good with words. And I began to understand why he is so considerate and thoughtful. What he lacks in words, he makes up for with actions that prove his love for me rather than proclaim it. So today it was very easy for me to know exactly what he was saying with every hand he lent me.

I appreciated my brother, David, being there with my parents to see me off today. David was eight years old when I was born. Although we are beginning to get closer to each other now, I proved to be a burden on him as a child. I was the baby that didn't know better, and he was still a child needing love and attention but from my Mom's point of view *was* old enough to know better. So when we fought I would always win; I would cry and he would get in trouble.

It wasn't until two months ago that I learned he resented me so much as a kid, even to the point of rarely coming home during his college years. There was a song we used to sing together that I thought at the time was fun because I liked to sing along with David, but it came out in a family conference in December that he was singing the song *at* me and not *with* me. It's a song by The Georgia Satellites called "Keep Your Hands to Yourself" and it goes something like, "Don't give me no lies and keep your hands to yourself!" To learn this really didn't bother me because it was something that happened in the past, and I was glad he was brave enough to come forward with it after all this time.

I wish David could accept himself for who he is rather than who he thinks he should be. I feel like he spends too much time searching for the answers in all the wrong places when he could find everything he is looking for right there within himself. I've come to realize that the grass really does always seem to be greener on the other side. You have to find a way to want what you have rather than to get what you want. So many people drive themselves crazy trying to get what they want and never realize the sea of things they already have to cherish.

I have realized so much of what I have just said in the past three months of my life, for that is when I met Abby.

I thought I had learned about love when I was in the fifth grade and "dating" a girl I met at the skating rink in Asheboro my grandparents owned. Julie was about three and a half years older than me but that didn't stop us from being together. One day, she asked me the question, "What does love mean to you?" I couldn't come up with an answer, so I questioned my friends and family. Everyone's answers, though often long and very different, included something to the effect that love meant loving someone. So I concluded there was no one answer to that question, and the only way to find love was to search for it and I would know when I found it.

I dated girls and tried to make this love thing happen, but it would always leave me heartbroken in the end. The more it happened, the more it hurt, and the further I would fall from wanting it anymore. I had been so heartbroken time and time again that I had given up on love and decided it just wasn't for me to find.

Just when the single life was beginning to treat me well, I find the love of my life in Abby. Eleven years it took me to answer the question my first girlfriend had asked me. And it was so simple—the answer is Abby! Her laughter, smile, eyes, generosity, freedom, intuition, imperfections, independence, persistence ... and I could go on and on about all the things I love about her. But most of all, she affects me in a way that completes me. I want to be a better person because of her. In fact, everything I have written up until this point and everything I will write from this point forward is because of her!

I have never imagined feeling as strong or loving anyone as much as I love Abby. I don't even think it's possible for anyone to love someone else as much as I love her and I submit that as a challenge to anyone out there who's in love or has ever been in love. If they could prove me wrong, or even come close to proving me wrong, I would be a very happy man to know that we as human beings are that capable.

This morning I wanted to hold her forever and was finding it almost impossible to say "Goodbye," but Lope Dog yelled at me because I was about to miss the bus. I didn't have time to think about not knowing how long it would be before I would get her first letter. And although I thought about the possibility of never seeing her again, I didn't have time to think about what it meant. We kissed, held each other close, and looked into each other's eyes one last time.

Then I quickly gave hugs and kisses and said my goodbyes to my family, and boarded the bus with my close friends, Ryan and Russ. Since we had a long journey ahead of us, we staged our gear in a separate seat so, all things considered, we could ride as comfortably as possible. Everyone waved as we waved in return. We watched them disappear as we disappeared.

Ryan, Russ and I talked for a few minutes about nothing, as we really didn't know what to say to each other. The bus got quiet; we were alone and able to relax in silence. It was great that everyone came to see us off, but at the same time it had almost been more than any of us could handle.

When Abby gave me my journal on my last trip to Charlotte, she asked me to write in it as often as I could so that when I returned home, she would be able to read it and feel what I was actually going through. In it, she had placed a letter and a list of instructions. She also gave me a book to read while I was gone—<u>Lord of the Rings</u>.

1-25-03

Eric J,

Thank you! As you lie beside me asleep all I can think about is how wonderful my life is now that you are a part of it. You have taught me more in the past two months about love, trust, security and humanity than my parents, teachers, and siblings have taught me in 23 years. If something horrible were to occur tomorrow and rip us apart (God forbid) I would still thank God for letting you be a part of my life at all.

"Thank you" does not even begin to describe the gratitude I feel for you being in my life. Thanks for accepting my imperfections, my insecurities, my mood swings and my goofiness. You are a great man—you are perfect right down to your smile, your blue eyes, and your beautiful face and body. Okay, so I am rambling now but I want you to understand all the gifts you have given to me in only two months time.

Thank you for being sweet and loving but strong and grounded. I Love You—I love you in a way that only love stories of the past can tell. People like you are the only reason that love stories exist at all. You make life worth living—your love makes everything from my past okay and bearable. I have always dreamt of a man just like you but I never dreamt that you could actually exist—you make me look forward to growing old and experiencing the rest of my life (with you by my side, of course). Thank you for today, yesterday, and tomorrow.

I Love You — Always,
Abby

Journal Instructions:

Write about the things you are experiencing, whether good or bad; the events of today will determine who you become tomorrow.

As we go through different experiences—day to day—our memories lose little bits of important details that affect our lives from here on out.

Write about your worries and frustrations.

No man can survive alone. The things we experience have also been experienced at a previous time—someone, somewhere has felt the things you are feeling right now.

Realize that I am doing the exact same thing you are doing.

Although we will be half a world away from one another, you are in my heart and brain and you will still be a big part of my life. You will be in my thoughts every second of every day, and you will hear every detail when you return.

If things get tough, remember and write down the good things that happened during that day.

If all else fails, know that I am thinking about you and waiting for you to return.

Chapter Four

"Hurry up and wait!" ... a saying that probably originated in the military and for good reason. I think it's caused by the lag time between commands beginning with the highest-ranking officer and ending with the lowest-ranking enlisted person. So it tends to have an accordion effect. Every unit has to be ready for a formation. The larger the level of formation, the more units there are that have to be ready. Everyone begins by being 15 minutes early for their immediate formation, which is 15 minutes early for the next formation, which is 15 minutes early for a larger formation; so forth and so on.

We arrived at the Marine Corps Air Station in Cherry Point where we were to board the plane. When we got off the buses we quickly grabbed our gear. Everything was done swiftly and silently, which seemed odd because usually everything is done in some sort of formation with someone yelling at us or singing cadence (which is really all the same).

Slowly, we began making our way into the hangar in an attempt to all fit in. Since there were more than 300 of us there along with all the gear we were carrying this was no easy task, but eventually we were successful and separated into our own platoons and companies.

Once we were set, we were to drop our gear in our place and sit on top of it. I struggled with this because there was a lot of hard work and preparation that went into fitting everything we were required to pack—items from the gear list as well as the personal items I had chosen to bring. If I sat on my gear, the personal items would get squashed and my pack itself would begin to fall apart. Then I would have to repack. It happens to everyone.

To avoid all of this, I chose to stand. Without asking, we mingled with our neighboring friends but we had to be very quiet in doing so since the officials were trying to get everyone through a single scale. If we got too loud then someone would yell at everyone else and we would all have to go back to our gear and be silent.

If we decided to smoke when the "Smoking" light was lit, we were allowed to do so outside the hangar. But the morning was still cold and damp, and none of us packed that many warm clothes because we were going to the desert where it was going to be hot. And if the thought of the hot desert was bad, the thought of wearing our MOPP gear in the hot desert was even worse—but that's another discussion altogether.

Many of us even shaved our bodies so we wouldn't get too hot. Now we were just trying to stay warm.

Then we were told to leave our gear where it was, go outside, and get in formation. Once that had been accomplished, we were instructed that before we would be allowed to board the plane, we were to take the two pills we were about to receive. One was red, and the other blue. (They actually looked like the gel tab pills from *The Matrix*.)

We were told that if anyone refused—as I was already doing—it would be a violation of multiple articles of the Uniform Code of Military Justice (UCMJ). The most obvious of the articles is "Disobeying a Direct Order from a Superior Commanding Officer." In addition, if we didn't get on the plane, then we would miss a movement. This constitutes a court martial and imprisonment.

They said the pills were for our own protection from the third world country we were going to. Our Platoon Sergeant, Commanding Officer, and Navy Corpsman were there to make sure we swallowed them. They stepped in front of us, watched us put the pills in our mouths, made us swallow, and then made us show them our empty mouths. Afterwards, they told us we couldn't give blood for another 5 years.

We went back inside, but our frustrations were only beginning. On at least three separate occasions we were told we were out of the order we needed to be in and had to shift one way or another. So as packed and crowded as it was in there, now we were trying to put on our gear and move to a new location within the hangar. It was like a real life sliding block puzzle.

Let me go over just exactly what my gear consisted of. I was wearing a Flak jacket that is supposed to be able to stop or slow a bullet which also has an ALICE (All-Purpose Lightweight Individual Carrying Equipment) pack attached to it that holds my two canteens of water, magazines for my M16, an E-Tool which is used for digging fighting holes, and a grenade pouch. I had a Molle pack (an acronym for a Modular Lightweight Load-carrying Equipment) strapped to my back which was packed full of items including a Kevlar helmet. Also attached to the Molle pack was my sleeping bag made up of a lightweight sack, a heavyweight sack, and a water-resistant cover. On top of the Molle pack was a two-man tent. Over my shoulders I had two full "bursting at the seams" sea bags, which are green cylinder-shaped bags about two ft. in diameter by about three ft. tall. And then, most importantly, I had my gas mask, M16 Assault Rifle, bayonet, and M240G Machine Gun.

Everyone's gear looks the same so when we moved, we had to move everything at once or it was lost. Since the shoulder straps of a Molle pack are generally very tight, we had to kneel over the pack, put our hands in the straps, and then lift the pack over our head upside down and allow the pack to fall into place on our back.

If no one is close, this is a very simple process. But trying to do it an area as crowded as the hangar was without hitting someone else with our pack, elbows, rifle, machine gun, etc., was almost impossible. Most of us were already tired and grumpy, so disagreements and arguments were inevitable. Now, imagine going through this entire process three times during this four-hour wait in the hangar, when we were rushed to get there.

Finally, we made our way through to the scales. To make sure our plane could carry our weight, we had to each be weighed individually with all our gear. On any given day, I weigh about 180 lbs.; now my weight was 344.

We then proceeded outside the hangar to walk to the United Airlines Boeing 747 that we were about to board. We could see the jet; it appeared to be just a few steps away.

I've flown on large jets before but had boarded from the convenience of an airport gate. I have also boarded smaller planes by having to walk up to them. But I had never walked up to and boarded a Boeing 747 from the ground!

The size of this jet was mind-boggling! The further I walked, the bigger it grew. I was astonished that something of this magnitude was able to fly at all, and we were about to weigh it down with all kinds of weaponry that I didn't think should even be allowed on non-military planes.

We all climbed what seemed to be mountains of stairs and then boarded. Seating went in accordance to rank, as things usually go in the military. I thought I was flying first-class, and in my mind it *was* first class! I was sitting at the front of the plane. I had a reclining chair with a footrest. And this was a full recliner, not a little lean-back. Ryan and I sat beside each other and had pop-up TV screens and our own headphones. Russ was a Lance Corporal so he wasn't yet an NCO, but he managed to sneak into the seat behind us.

At that point, my day was looking way up! The flight attendants were going to be some of the last females and civilians we would see for quite some time. It felt pretty cool just to ask for a pillow from an attendant when I had an assault rifle and a machine gun in my lap that were almost as long as me.

This was a really exciting time for us because again, we had not known what to expect. We thought we might be sitting on the floor of a cargo plane or something. I had to tell somebody about this because it was just too cool not to be able to tell someone about. I had refused to leave my cell phone behind, and it was sitting in my pocket in case I ever had the chance to use it. I powered it up and called Mom and Dad as well as Abby to tell them how excited I was and how well we were being treated on the plane.

Before we took off, we got word we would be flying seven hours to Frankfurt, Germany. Once we were cruising, Ryan, Russ and I talked for a little while about what we thought we were going to do. I was excited about going to Germany because I had never been there or for that matter, to any country outside the United States other than Mexico. I was even excited about having to fly for seven hours on this plane.

As a little time passed, Ryan and Russ began writing letters home to their girlfriends, and I began playing with the TV. There was a play list of movies I could watch, music channels I could listen to, and even a GPS map showing the plane's location and estimated time of arrival, distance traveled, air speed, etc. Then Ryan and Russ fell asleep, but I was wide-awake. It had been my plan to stay awake during the entire flight, and that is what I did.

I ate what was to be my last American meal, and I must say it was delicious and nicely presented. I was also able to watch two movies, listen to two full-length music channels (after an hour or so they replay all over again), and I was able to write my first entry into the journal Abby had given me. I could not get her off my mind, and was already beginning to miss her.

We landed in Frankfurt in the dark at about 6:30 the next morning, so I wasn't able to see much other than the usual lights. Lights don't look that different from place to place unless you're flying into Vegas. It was kind of disappointing, but I was O.K. with it because I knew we would be seeing the airport and also it might be daylight when we left.

We exited the jet and left all of our things there. It felt really strange leaving our weapons because a Marine never leaves his rifle behind. Even stranger than that was the walk to the airport pier without my rifle. I felt like I was forgetting something ... that was because we all had to learn the hard way back in boot camp that we never leave our rifle behind or even drop our rifle without following it to the ground.

If we dropped our rifle in boot camp, we would hear the Drill Instructor yell, "Follow it!" Then we would drop down, rest our rifle on the back of our hands, and do push-ups until we were told we could stop. This was all while the Instructor pushed the rifle down with his boot on our hands. If we were to forget our rifle, a burial ceremony would be held. We had to dig a hole in the sand, bury the rifle, and then of course dig it back out and clean the entire weapon with basic cleaning gear that resembles a toothbrush. And if all of this wasn't bad enough, it happened in front of the seventy or so other recruits in the platoon. (We were called "recruits" in boot camp because we had not yet earned any rank or title.)

On the flip side of this punishment is laughing at someone who has to go through it. That's where I would always get in trouble because honestly this scenario was hilarious to watch and I couldn't help but grin or burst out laughing. Then I'd have to join them, making it not so funny.

When we were inside the Air Force pier of the Frankfurt airport, we learned we were restricted to that area. So then I really was disappointed; it looked like I might not get to see any Germans. But we had an hour layover and it would be daylight when we took off again, so I still had hope.

We were supposed to be there until 0745 and then fly directly to Kuwait, but someone didn't plan accordingly and we didn't get to board the jet until 0900. (I never did get to see anyone I thought might be German.) Apparently there was another unit flying into Kuwait City at the same time, and the airport wasn't big enough to handle two jets of our size landing simultaneously.

Well, we sat in the jet for another hour or so waiting to take off, but nothing was happening and we were all getting a bit antsy. Finally the Australian pilot who spoke with an accent came over the intercom and apologized for the delay. He explained, "The weight of the paperwork had to have equaled the size of the plane," which lightened the mood just a little.

We took off, and I could see the Swiss Alps below me. It has always been a dream of mine to go snowboarding in the Swiss Alps and one day I'm going to do it. We also flew over the Pyramids, but I really don't care to go visit those.

I had planned on staying awake for this four-hour flight, but my emotions finally caught up with me. Luckily, Ryan and Russ had already fallen asleep; the last thing I needed was for them to pick on me for crying. During the movie I was watching, *Stewart Little,* there was a scene where the little mouse saved a little bird. It was just enough to bring a tear to my eye, which was just enough for me to realize how much leaving everyone hurt.

I didn't know where I was going and I didn't know if I was coming home. I didn't know if I was going to be able to write home or receive mail because it might jeopardize our security by giving away our position. I didn't know if Mom was going to be able to make it in health or if she was going to lose it mentally. I didn't know if Abby was going to be able to handle having a boyfriend that she couldn't talk to or even know if I was alive. I asked myself if I would wait if the roles were reversed, but I couldn't fully convince myself I believed my answer.

At this point, I began to cry and I remember asking myself so many questions and not having any answers. It is the fear of the unknown that causes the sense of helplessness I was feeling. This was my "cleansing of the soul." After all those months of build-up, I finally knew how I felt.

I tried to hold back the tears, but the harder I tried the louder I cried so I just let it happen. The more I cried the more I wanted to cry. Crying felt good, so I thought about other things I'd been upset about in the past and cried about them, too. Then I tried to convince myself that everything was OK … or at least *would be* OK. I tried to cling to something as if I was trying to reach out to someone, but no one was there.

Eventually my tears subsided, and I fell asleep.

Chapter Five

I woke in time to see Kuwait City from the air as we approached the airport. It was much more developed than I had pictured it to be, but I could see it was very poor. Most of the roads were sand; there were very few paved highways. It was difficult to tell, but it looked as if most of the Kuwaitis drove sub-par SUVs and the government drove nice luxury vehicles.

As soon as we landed, we swiftly exited the jet and immediately boarded Kuwaiti buses. With the interior lights off and the curtains closed, the buses took us to a remote location within the airport complex. Then accountability was taken of everyone and 7-ton trucks were loaded with our Molle packs and sea bags. We received forty security rounds of ammunition, as we would be traveling in Threat-con Charlie (Threat condition "C"— meaning terrorist threat is high). We were briefed that we would be traveling north for eighty to ninety minutes to our destination of Camp Coyote. It seemed like they had it all figured out and that things were really coming together now. But two hours into the trip, I began to have doubts.

We pulled over at a camp on the side of the road so I asked, "Are we taking a break or what?"

"This is it," Potato Head answered and we got off the bus to again take accountability. (We had given Potato Head his name because that's what he looks like—he has a really big head, a small patch of hair on top, and narrowly-spaced beady eyes.) After thirty minutes of standing there alongside the road with no one knowing what was going on, my doubts were proven true. We were at the wrong place.

We got back on the buses and started again. After another thirty minutes or so, some of our entourage—Light Armored Reconnaissance (LAR)—reached their destination and got off the buses. So Ryan, Russ and I got off our bus too, in order to find a bus less crowded. Now it was just us and a couple of others, and we all had two seats each. This was not only great because we had some breathing room, but the trip was about to get more interesting.

We were traveling on sand roads with no traffic dividers or road signs. The Kuwaiti bus drivers must have thought they were racing the Baja 1000. We were three buses wide going down the straightaway, and then we'd all dive into position going through the turns and slide sideways on the sand.

Ryan, Russ and I were the race commentators and it was so much fun. We made another stop but it was the wrong place once more. So we made a dash to the first turn, and the race was on again! Of course by this time the curtains were wide open so we could announce the race. Our fears of terrorist threat were out the window; it was race time, folks!

Unfortunately, all good things must come to an end so at 0230 Sunday morning, we arrived at Camp Matilda. This was apparently our new destination, and was supposedly located about twenty-nine miles south of the Iraqi border near Al Basra. They said that in a month we were going to be at the line of departure. I wasn't sure what that meant exactly, but I thought it had something to do with being at the furthermost point before invading Iraq. I equated it to the "point of no return."

We tried to sleep for about three hours after we arrived, but it was impossible. The 7-tons with all our gear went to the wrong place, so everyone shivered in the cold all night instead of sleeping in warm clothes or a warm sleeping bag. I don't think any of us realized how cold it was going to be there for those first few weeks.

But fortunately a little later that morning most of us were able to get our gear and we were fast asleep by 1000 hours. At 1800 hours, we had to get up and set up our two-man tents within the larger circus-type tents we were staying in. Apparently the United States paid the Kuwaiti government entirely too much money to set up these circus tents for us. But in actuality it was good because the two-man tents don't do a great job of holding back the elements. Having the two-man tents inside the circus tents also gave us a sense of privacy, not to mention the "accountability" factor ... when we know where someone belongs, we know who is there and who is not.

My tent "roommate" was to be Cpl Kammerer, or "Kammerererer" as we call him since he has so many "er's" at the end of his name. He was my assistant gunner, or A-Gunner, on the M240G Machine Gun. We didn't talk that much, although he's really a decent guy. Ryan and Leonard were my neighbors so I would have someone to talk to and hang out with.

After we got all of our two-man tents set up, we had a formation to "pass word." It was then we learned that Field Mess had evening chow on the way for us and that we would be served hot chow that evening. We all enjoyed eating hot spaghetti and drinking juice at a circus tent that was set up with tables similar to an actual chow hall.

Until that meal, the only things we'd had to eat since our dinner on the first flight leaving the States were MREs (Meals Ready to Eat). MREs come in all different menus and all things considered, I think they actually taste pretty good. I just can't figure out how (and it worries me just a little) they can make the beef and chicken last seven years before expiring. The MREs also contained flavored sugars so we could make lime, cherry, or grape flavored water. During field training operations back in the States, these mixes would just get thrown away because the only containers we had to mix drinks in were our canteens. Once a canteen is contaminated with a flavor, water put into that canteen would always have a hint of that flavor. In plastic canteens, this hint of flavor tastes like mold or mildew and of course that's the last thing we wanted to drink. But when they found we had no cold drinking water available, the Kuwaitis hooked us up with 1.5L bottles of water. So now we could mix worry free.

For the remainder of the day, Ryan, Russ and I would go exploring Camp Matilda. We found trailers that had three showers and three sinks each, and there were six trailers close to our tent. It was a huge relief to know we were going to be able to take decent showers, and they even had warm water for those in the front of the line. There were also nearly a hundred Port-a-Johns scattered around the camp, and the Kuwaitis were to clean and restock them once a week—another huge relief.

That night, I was writing a letter home to Abby and realized I had failed in the rush before I left to arrange for flowers to be delivered to her on Valentine's Day. Last year I was dumped on Valentine's Day and I had told myself I would never observe the day again. But of course all this changed because of Abby. I felt irresponsible for not planning ahead, so I also wrote a letter home to Mom to see if she could help cover for me. In my letter to Mom and Dad, I tried to include whatever information I felt I could relating to our location and plans ...

> *"I can't tell you exactly where we are in Kuwait, but it is about the distance from our old home to our new home—actually about 1/3 less, from a border. Remember that guy Basra? I can't remember his first name, but for some reason Al comes to mind. Well, if you were to look at a map, you would find him above us about the distance I told you. And from hearsay, I think we are going to visit him in about a month or so ..."*

And I asked Mom if she would send Abby 12 red roses, 2 white roses, and 29 Hershey's kisses, along with a love note I enclosed to go along with the flowers and candy.

My letter to Abby, along with telling her how much I missed her and how much she meant to me, recounted the events of our trip and our Camp Matilda living conditions. And, since we had all become close friends, I thought she might be interested in Ryan, Russ and his girlfriend Kate.

> *"Russ is writing to Kate, and Ryan and Leonard are playing tunes on their harmonicas. I told you that Russ and Kate had gotten serious but I don't think I realized just how serious it is. He bought her a promise ring and they talked about getting married. Supposedly Kate wanted to get married before he left, but Russ told her that if she still felt that way when he returned, then they would."*

At 2200 hours, the lights in the tents went out. But it was hard to sleep; the Firewatch was walking circles around the tent and with the plywood flooring, I could feel their footsteps as they walked by. They were also carrying flashlights they had to keep on to avoid tripping over anyone, so I tried to keep something across my face to block out their light.

At 0300, I was at the point of not being able to take it anymore when I heard Ryan in the tent next to me talking to himself as he does when he's upset. So I rummaged through my pack to find an MRE and crawled into Ryan's tent to hang out with him and Leonard. We talked a little and picked through the snack parts of the MRE.

At 0400, I went back to my tent with Kammererer and tried to go to sleep. I brought a mini-disc player with me and made a few MDs before I left that had about fifty songs each. I listened to my country music MD and never slept at all that night. But I kept thinking about one song I did not have—a George Strait song that had been playing over and over again in my mind for the past couple of days. With my flashlight in one hand and pen in the other, I wrote Abby and told her about it. The title is "Carrying Your Love with Me." I told her how I had changed some of the lyrics to fit our situation, and how the perfectly the verses "You're right there in everything I do," and "All I care about is getting back to you," expressed my own feelings for her.

Somewhat surprisingly, the next day was a good one. It was lunchtime when I made my journal entry and we had gone that entire morning without getting in trouble. We also managed to pose as a working party and steal a case of water to bring back to our tent. (They told us that water supplies will be scarce for the first week or so; we had to do what we had to do.)

After all of this productivity, Ryan and I ran across a group of Kuwaitis who were there working on our Camp, and we managed to get our picture taken with them. It was our first meeting with anyone from their country, and our encounter with each other seemed to make their day as well as Ryan's and mine!

Things continued to go well when I had Chicken Tetrazzini for lunch. That's really just a fancy name for chicken and noodles, but regardless, it's one of my favorites. I'm typically a beef guy with little or no desire for chicken, but with MREs, beware of the beef!

That evening, we ate hot chow again. It was nice having a day go by that we weren't being yelled at. Unfortunately, there was a rumor that night that Iraq was planning an attack. We didn't think the rumor was true, but at the same time, things were tense.

Chapter Six

Three days have come and gone now, and I feel like I've done nothing. I'm going through the initial shock of being here and being bored. But when I'm fighting feeling homesick, there is nothing worse than boredom because I don't have anything to distract me from thinking about how miserable I am.

Ryan tells me I should come out of my tent and hang out with everyone. Sometimes I wish I could, but then there's a part of me that says not to because I'll do nothing but bring everyone else down with me.

I used my Smallpox vaccination the other day as an excuse for not coming out of my tent. I said it had me feeling bad, which was partly true. I didn't feel well that morning but was OK after a nap—at least physically.

I wrote a letter to Abby, but even that wasn't as emotionally pleasing as I hoped it would be ...

Feb. 13

Abby,

What's up, Baby? I just woke up. You'll never believe that I slept from 10:00 AM yesterday to 6:00 AM today! My Smallpox vaccine finally took effect and I felt like shit. So I stayed in my tent all day long. I was able to read another chapter in my book. At 3:00 this morning, a thunderstorm passed through and woke me. But at 4:00, my flashlight died and I went back to sleep. Sorry that I don't have much to tell you, but my days have mostly consisted of sleep.

We had classes on how to handle POWs as well as another class on NBC. Then Ryan, Russ and I had to give a class on the 240G Machine Gun. We were also able to talk our platoon sergeant into giving us an Intel brief. I wish I could tell you what our plans are but I can't. I'm about to write that down elsewhere. I can tell you that I doubt I will ever see any action. I think they are leaning toward us dealing with the POWs.

I wish the mail would hurry and get here. I hate not being able to hear from you. Tomorrow is Valentine's Day and it sure would be nice to get some mail but I doubt it *will happen. I need you so bad right now. They say that if you love someone to let them go, and if they come back, then that's how you know. But I think they are wrong. I love you and need you and I don't ever want to let you go ..."*

All I want is to be with her right now or just be able to hear from her. In my previous letters to her, I couldn't wait to write and tell her about what I was going through. But that morning's letter was almost dreadful.

I'm not exactly sure why I felt this way. Maybe it was because I had done nothing the entire day and was embarrassed to tell her. I felt like I should have all these amazing stories to tell, or that maybe she's expecting to hear me tell amazing stories.

Maybe it's because I have written so many letters already but I still haven't received anything from her or gotten any sign that she even cares about me. To add to that, Marines sometimes have a way of making secure men feel very insecure about themselves. And we hear stories of girlfriends and wives of Marines sending pictures or videos in some cases of them having sex with others. While I really don't believe anything like that would ever happen to me, I can't get the doubts completely out of my mind.

Could I just be losing hope? It's only been a week but it feels like an eternity. I'm probably just grumpy and want to be alone. Whatever it is, I want to go home.

To take my mind off Abby, I wrote a letter to my best friends, Aaron and Patrick. But I doubted that either one of them would actually write me back. Not because they don't care; they're just guys and don't typically write letters to other guys.

The Marine Corps does a number of things to keep their troops busy. One of those things is to have one Marine give a group of Marines a "Period of Instruction." They always start the same, no matter how many hundreds of them you get ... "For those of you who do not know me, I'm Sgt so-and-so and your next period of instruction will be on blankity-blank." The reason I point this out is because everyone knows who the guy is giving the class and quite often the people receiving the class are more familiar with the topic being covered than the one giving it.

Our first period of instruction at Camp Matilda was on the M203 Grenade Launcher. Clearly this class was only to keep us busy since every Marine in the world learns the basics of an M203 Grenade Launcher in boot camp and further in Marine Combat Training. So the newer the Marine, the better they know how to use it because the information is still fresh. But now we had a Sergeant giving a class on this who hadn't used one in ten years. We were all dumber at the end of this period of instruction because the instructor wasn't up to speed on what he was teaching.

Our second period of instruction was on the M240G Machine Gun, taught by Recon Ricky himself. Recon Ricky earned his name not because he was Special Forces, but because he was not. He was your typical "one-up" kid. Anything you had done, he had done better and he had the "stories" to back up his experiences. Other than that, he's an O.K. guy but he had already pissed me off that day. He had been yelling and cursing at everyone for no good reason, so I decided to challenge him. While he was giving his period of instruction on the M240G, he started speaking from his "experiences" more than about the gun itself. So I asked questions about the gun that he didn't know the answer to, and corrected him at every opportunity. When it was clear he was completely confused, I reviewed for everyone the basics of the weapon and how a 2-man team operates it.

To waste even more time, we were given a period of instruction on a sand table. It was just a way of showing a plan of attack like they did in *Robin Hood*. And then we had a class on how to apprehend an enemy prisoner of war (EPW). This was going to be a huge job for the support companies, because the assault units were taking no prisoners.

After these periods, we were able to talk Potato Head into letting us in on our planned course of action.

We were to move all of our troops up to the border. Tanks were to have the biggest role, since they have the greatest firepower. This was only a bluff, though, and we were to move back out of sight and repeat a couple of times in an attempt to provoke Iraq to take action. Finally, we were to roll through, destroying the main targets and bypassing smaller ones. Our entire regiment was to be combat ready by February 20. The Kuwaiti schools were scheduled to close by mid-March, at which time everything was to begin.

No one really knew if they could believe Potato Head because he was known for making up stories. At the same time, he was also known to be the first Staff NCO to give us the scoop. So if anyone knew what was going on and would tell us, it would be him.

In a lighter vein ... later that evening we were in a thunderstorm, complete with rain. This was unusual, being in the middle of a desert. You know what else was strange? There were dogs that barked at night that made us feel like we were camping out in the middle of a city rather than a desert with not a town in sight. I heard someone say they had seen a camel the other day, but I've yet to see one. I haven't seen a scorpion either. I guess I'm a little disappointed.

The next day would be Valentine's Day. I sat down and wrote a letter home to Mom and Dad, and asked them if they could send me some baby wipes. If and when we go north there won't be any showers, and we can use baby wipes to take hygiene showers.

And of course I wrote a letter home to Abby. It was just so heartbreaking not to be able to spend Valentine's Day with her. I even wrote her a poem to try to express my feelings.

Unfortunately, Valentine's Day came and there was still no mail received by anyone. I remembered that the February birthdays of my cousin, Darren, and my nephew, Josh, had also passed.

Darren is about eighteen months younger than me. We lived just a short walk through the woods from each other with Grandma's house in between. We all live on land that has been in the family since the 1850's and had dirt bikes, so we were always together. Even as youngsters, we were VIPs at the skating rink my grandparents owned, and in our small town Jones' Skating Rink was the place to be every Friday and Saturday night. We had the newest skates and the best wheels. We had been skating longer than any of the older kids that came. Everyone knew us, and we liked to show up about an hour late every night in order to make our appearance.

When I turned 16 and got my driver's license, Darren and I began to lose touch. I quit going to the skating rink, and he wasn't allowed to ride in the car with me. Eventually I would leave the small town heading off to the military, but he never did. He turned 21 last week.

Josh is my nephew, my sister Kim's son. I was only 10 years old at the time Josh was born. I thought his coming along was the coolest thing that had ever happened. I treated him as if he were my own. I'd had pets before, and didn't treat him any different. I would carry him around with me and show him off to my friends, especially the girls. They loved him and admired me for having him with me, and I loved every minute of it. Josh and I got along great; I was one of the few who could take him from crying to laughing so quickly and easily.

I had a girlfriend at the time who was several years older than me. When he came along, I honestly believed I was ready to have a child of my own in my life. The only thing I thought I couldn't do was change diapers. I didn't think about the fact that I couldn't work and make money either, so I guess it's a good thing it never worked out for me.

But back to Valentine's Day. It was both good and not so good. It was good because I wrote my first poem to Abby and sent it to her as a card. I knew she wouldn't get it for a while, but at least when she did she would see the date on it. I was pleased with myself because I had never been one for writing poems. I guess I had just never had the right inspiration. Ryan liked it too, so he copied it and sent it to his girlfriend, Bridget.

On the not-so-good side was the ridicule I caught from Ryan and Russ over all things ... a haircut! It had been three weeks since my last haircut, so I was overdue. Every Marine keeps a fresh haircut and is required to have one at least once a week. I decided that since I wasn't trying to impress anyone out here I would shave it all off, so I did. A little unhappy with it, I went one step further by taking a straight razor to it.

I expected to get hell from Ryan and Russ when they saw it, but they just wouldn't let it go. Russ told me that it might work if I looked like Vin Diesel but I was way too small and I needed to get back in the gym. They were calling me names like Powder and asking how my Chemo treatments were going. And then they started calling other people over to join in on the fun.

If it had lasted just a little while and stayed between us then I would have been fine with it. I had expected it, and their comments wouldn't have been a big deal. But in my current state of mind, I could only take so much; I finally decided it would be best for me to go away so I retired to my tent for the rest of the evening.

Ryan is one of those people who don't seem to have a care in the world. But when there's a bandwagon to jump on, he's right there. The only reason he was hounding me was because of Russ.

When we left for Iraq, Russ was serious about a girl named Kate. Kate and I had met some time ago in Charlotte and had become more than friends. She had wanted a relationship, but I didn't like her like that. I introduced her to Russ, and they hit it off pretty good. The problem Russ had with this was I had Kate before he did and although I tried to help him with that, he never could seem to get completely beyond it.

The next day when I saw Russ, he started in on me again. Ryan started laughing, and then it was both of them. It seemed to be much more than good-natured ribbing. I had the feeling that for Russ the jabs had to do with something other than my haircut. It seemed personal. I got a little angry, and told them I didn't want to say anything that any of us would regret. All I would have had to do was say something about Kate and feelings would have been hurt. Russ quickly got the picture, and like usual, Ryan followed suit. It was finally dropped.

Now my hint of a cold that I thought was just the Smallpox vaccine had worsened. Medical had nothing to give me, but told me it was the Smallpox still in my system. They told me to drink more water and it would be gone in a day or two.

February 17

Abby,

Hey Baby! Sorry I have not written you in a couple of days but I've been sick, and still am. They tell me that it is normal to feel this way because of Smallpox, but I've been like this for 5 days now and feel like I have strep throat. I mostly spend my days here in the tent just sleeping and reading. I'm up to Chapter 3 of Book 2 in my book. It finally got good yesterday and I couldn't put it down.

Our ship is supposed to come in today, so I get to go to the port to drive back some Hummers and stuff. I'm looking forward to it, but just wish I felt better. When our equipment gets here, things will pick up and I won't have much free time.

Of course we haven't gotten mail yet. Cigarettes are going for $10 a pack now and will increase when we move onto the line of departure ...

Three days later I was supposed to go to the port to pick up some 7-ton trucks and HMMWVs (or Humvees). The letters stand for High Mobility Multi-Wheeled Vehicle. I wasn't able to go because I was still sick. This was very disappointing; I had been looking forward to the trip and getting to do something other than being inside the restrictions of Camp Matilda. I went to Medical again only to be told the same thing they told me three days earlier. They told me they still did not have any medicine to give me, but after a few choice words I was able to acquire two decongestants.

I finally felt better the next day and was able to go to the port. It was pretty cool just to get away for a while. We rode in the back of open 7-tons on the way down which was great; although our sight was limited because we were riding backwards, we could still see out.

At the port, we found the living conditions for the Army, Navy and Brits stationed there were quite different from ours at Camp Matilda. They lived in air-conditioned tents. They had a chow hall that served four hot meals a day, soft drinks in refrigerated coolers, a PX, and a phone center. I went to find the phone center to call home, but they said I couldn't. I had to have a code issued to me or a calling card, and I had neither.

I was able to have lunch in the chow hall. It was really nice having a hot meal and a cold soft drink. I normally don't drink soft drinks and actually prefer water, but when something is taken away, it's usually nice to get it back.

On the trip back I was able to ride with Lope Dog in the Humvee, which was again great because I was able to see so many things. A donkey, lots of herds of sheep complete with shepherds, and even some camels. There was actually a herd of camels in front of a Toyota dealership!

We passed two bazaar cities along the way that looked as if the entire cities were being built at once. There were too many buildings to count, but not a single building was finished. Instead of constructing one building at a time, they were building one city at a time.

The Kuwaitis smiled and waved as we passed them and seemed to jump for joy when we waved back. We had a police escort for our convoy on the way back. At all the intersections, people were waving from their cars and giving us "thumbs up" as if we were in a parade. I tried to take a picture of one of the road signs that had at the top Arabic writing but across the bottom read, "God Bless the U. S. Troops," but at 60 MPH and with a disposable camera it was kind of difficult.

When we got back to Camp Matilda, everyone wanted my Sprite but I didn't trade. It had now been three months that Abby and I had been together so I wrote her a Happy Anniversary letter. And I wrote home to Mom and Dad.

February 22

Abby,

Happy Anniversary! So three months have finally passed. I hope that you enjoyed the last letter I sent you. Don't worry ... I didn't let anyone read it or copy it. Ryan copied the poem I wrote you and sent it to Bridget as "our" poem, yet the only part he had in it was copying it to his letter.

... We finally have our vehicles and equipment here, so we've been working late. I am also getting over my cold, except for a little congestion. I didn't get to go to the port the day I told you because I was sick, but things worked out for the next day and I was able to go. The Kuwaitis love us; we had a police escort for the convoy on the way back. Everyone at the intersections was waving, giving us thumbs up, jumping up and down and laughing if we waved back to them. I felt kind of important for once.

I wish I could tell you what is going on but I'm afraid this letter won't reach you if I do. I still haven't gotten any mail yet—none of us have except Russ. He got a package from his Mom but no one can figure out how. It was postmarked Feb. 12, and I know people mailed stuff before then ...

... Try not to worry about this movement that's coming up. I'll be all right and home before you know it. We have a pool going on when we will go home. I bet Aug. 21, but the majority say June or July...

February 22

Mom and Dad,

Hey! Well, I'm finally over my cold! Will you do me a favor and wish Darren and Josh a Happy Birthday? I kind of forgot about them. Dad, by the time you get this, you'll be what? 55? Wow! I'm just kidding ... I wish I could be there for you but know I am thinking of you; HAPPY BIRTHDAY! You know, for people that I go out with on their 21st birthday, we make sure they take 21 shots of liquor. Can you handle 55? Maybe we can change yours to 55 drops? Ha ha ... I somehow doubt you'll do either ...

... Take care and I'll write again soon. Oh, I almost forgot—I'm about 400 pages into my 1,000 page Lord of the Rings book ....

Chapter Seven

The next day all the vehicles arrived, and we worked toward getting each one of them combat effective before 1600. Of course, that didn't happen. But for a change I actually wanted to work just so I'd have something to do despite the sandstorm.

A sandstorm is basically a windstorm in the desert that stirs up sand. Although they usually don't last very long, the storms sometimes last for hours. The wind pelts us with thousands of tiny grains of sand that irritate and blister our skin. The sand covers our bodies and hair. It gets in our eyes, nose and mouth. It makes breathing difficult, and is especially uncomfortable for my eyes because I wear contacts. Sandstorms can also become dangerous in unfamiliar territory; visibility is often reduced to ten feet or less and there is a risk of quickly becoming disoriented and wandering off.

Anyway, we had put up our maintenance tent—a big, white rectangular tent about 30' wide, 40' long, and 20' high with two large doors on either side—so we could pull our Tracks into it for maintenance. The sandstorms didn't have much effect inside the tent as long as the doors were all buttoned up nice and tight. The tent was connected to a generator; it was very well lit and we could work around the clock. It was also light restrictive so it didn't reveal our position.

I actually tried to work, but ended up being kicked out of six different areas. They would say they already had enough help when they really didn't. It was just that they didn't want Ryan, Russ and me hanging out together; we were known as troublemakers.

But when I stopped working, a Staff NCO would yell at me to get to work or find something to do. So I did, but then would get kicked out again because Ryan or Russ was near. I eventually gave up trying to help and went back to my old "skater" ways.

Skating, or being a skater, is not as easy as it sounds. It's actually working to avoid work. It goes hand in hand with the "Out of sight, out of mind" philosophy. Ryan, Russ and I were professionals at it. No one could yell at us when we were out of sight; we rarely got caught but if we did we usually had an excuse ready that would cover us.

It's almost sad that I skate so much. I think back to my senior year of high school when I spoke to the Marine Corps Recruiter. I was almost ready to enlist, but needed to think about it a little more and had to ask my parents. Mom and Dad didn't approve, making me want to join even more.

But knowing how much I hated school, I decided that college would probably not work out for me and would end up being a waste of time and money. I knew I had some growing up to do, and I also knew the Marine Corps would help my credentials as well as pay for school so I wouldn't have to ask my parents for the money.

I had taken the military's ASVAB test one day in order to get out of my high school English class. My teacher had asked for volunteers to take this test and said what we scored on it did not matter. The test also excused me from a class I usually slept through, so it seemed like a win/win situation to me. What I did not know was that the score I made would determine how many and how often military recruiters would call, and how determined they were to recruit me.

For one reason or another, I have always been a good test taker. I would like to believe it's because I have good common sense. But I think it's because I always paid close attention in the test-taking classes we were given. It's another example of education I never agreed with yet embraced. In my mind I had two choices. I could thoroughly learn the material being taught, or I could learn how to take tests and not have to worry so much about the material. I chose to learn how to take the tests, and got through school with minimal effort.

This came back to bite me when I took the ASVAB. Apparently the score I made qualified me to do any job the military had to offer, and prompted numerous recruiting calls. They weren't going to give up no matter how many times Mom hung up on them. One evening, I answered the call from a Marine Corps Recruiter, and spoke to him at length mainly out of curiosity just to see what he might say. I even allowed him to set up an appointment with me that at the time I felt sure I would not make. But this guy knew what he was doing. He challenged me by telling me he knew my type; I was only wasting his time and probably wouldn't even bother to make it to our appointment. Well, I sure showed him.

I enlisted in the Military Intelligence occupational field for four years, and was required to take an additional test for this occupation called a DLAB test (short for Defense Language Aptitude Battery). One of the first questions asked me how strongly I felt about learning a foreign language and to answer a number from "1" to "5," with "5" representing that I strongly agreed. I answered a "1" indicating that I strongly *disagreed* with learning a foreign language ... not wanting to go to college was why I enlisted! In no way did I want to go to this type of school.

I went to boot camp knowing that I was going to have a job in the intelligence field but I didn't know what. Everyone goes to boot camp knowing what field they're going to be in, unless they sign up as open. But few people, if any, know what specific job they will get. Near graduation, everyone finds out what MOS school they'll get orders to for job training.

When I opened my orders, I couldn't believe my eyes. I was going to the Presidio of Monterey in California where I would attend school at the Defense Language Institute. If there was one job in the Marine Corps I did not want, this was it. Being that I did not excel in English, it was going to be very difficult for me to learn foreign languages that rely on our own system to teach other systems.

When I got to DLI, I had to take a test to show my learning patterns so they could place me in the language best suited for my ability. I qualified to learn the Russian language—a Class III difficulty course that condensed eight and a half years of college Russian into one year. The Spanish language that I had hoped for was only six months long. And if that wasn't enough school, after I graduated from DLI, I was to go to Goodfellow Air Force Base in Texas for six additional months to learn the top-secret portion of my job. Because of all this instruction, my enlistment had to be extended by a year to five years.

Four months into my Russian school, the Marine Corps Detachment I was with did a surprise "health and welfare" inspection of the barracks. These inspections are not designed for our health and welfare; they are designed to search the barracks for any illegal substances or contraband.

Before the inspector (who was my platoon sergeant) entered the room, he said, "Now is the time for amnesty." He explained that if I had anything illegal or should not have, I could tell him then and not be punished. I really didn't think I had anything, but then remembered I had a commemorative bottle of White Zinfandel house wine from the Marine Corps Memorial Hotel in San Francisco, as I collect commemorative items. I was only twenty years old at the time so technically because it contained alcohol I was not supposed to have this item. Barracks regulations prohibit anyone under the age of twenty-one from possessing alcohol.

I told my platoon sergeant about it and went inside and got it for him. He told me he would have to hang on to it for the time being, but that I would be able to get it back soon.

A couple of weeks later, I was told to report to the Company Gunnery Sergeant who happened to be an interrogator translator of Cuban descent. This was one guy I did not want to have to answer to. He asked me about the wine, and I told him I had kept it as a collectible and was not drinking the alcohol. He advised me, "Tell that to the Man!"

The day before my scheduled promotion to Lance Corporal, the Detachment Master Gunnery Sergeant read me my rights before I went in front of the Commanding Officer for my Non-Judicial Punishment. At that point, I told Master Guns that someone else had purchased the bottle for me when in actuality I had purchased it myself and was never carded. I requested legal counsel after telling him this and went to speak to a JAG Officer.

The JAG Officer advised me that after lying to Master Guns, I should exercise my constitutional right afforded by the Fifth Amendment and just remain silent.

But no matter how respectfully I requested, the Commanding Officer was not at all happy about me remaining silent and not telling him how I got the bottle of wine. He said that in addition to violating barracks regulations and performing acts that were unbecoming of a Marine, I also had a character flaw. Although a character flaw is a matter of personal opinion and not chargeable under the Uniform Code of Military Justice, his opinion hurt me the most. He charged me with breaking two articles of the UCMJ, demoted me to the rank of Private, and kicked me out of his school. Being kicked out of school also meant I would not be able to get the top-secret security clearance that would have benefited me.

This was his way of maxing me out for my character flaw, when generally the alcohol violation would have just been a slap on the wrist. Prior to this, I had decided the Marine Corps was going to be my career. I was preparing a package to be considered for admission into the Naval Academy, as I wanted to become a commissioned officer. After the Marine Corps, I wanted to work for the federal government, maybe in the FBI or the CIA.

I was ready to devote my life to the military and my country, and could not have been more proud of it. Unfortunately my dream had been shattered. I would never be able to attend the Naval Academy or become a commissioned officer due to my mistake and this one Major's opinion, yet remained committed to the additional year of service. So, for the next four years of my life, I was in effect sentenced to a job and accompanying working environment that would have been unimaginable to me at the time I made the decision to enlist.

Due to the severity of my punishment and the JAG Officer's advice, I did file an appeal. But when I followed up after several months to see why I had received no response regarding the outcome, I learned the appeal paperwork had somehow been lost. I decided that pursuing the matter any further would most likely be pointless.

Chapter Eight

The only work I did the next day was to help put up camy netting. Camouflage netting scatters satellite imagery so it goes over vehicles, tents and equipment to help hide our position.

That night, I began to get lonely again. It seemed like an eternity since I had been with Abby. I decided to write her an erotic letter. I told her in as many descriptive words possible what I wished I was doing to her right then. My imagination ran wild and it actually began to get quite humorous yet still intrigued me to write in more and more detail. I hoped she would enjoy it.

Over the next week, things got pretty crazy. It was our understanding that we were to advance north to the dispersal area on February 27th. We were working around the clock in order to meet combat effectiveness deadlines. We had different crews working in shifts so that two crews worked while one rested. But as usual, the deadlines weren't met and had to be pushed back repeatedly.

On February 24th, I got my first piece of mail from home. It was a Valentine's Day card from Mom, but it was actually kind of depressing. I couldn't believe that instead of writing me a letter, Mom would just send me a card. It was a sweet card and I did appreciate it; it was just that it wasn't what I needed after three weeks of no contact. But in the card she wrote something that didn't make sense to me. It was then I realized she had written a letter before sending the card and that I had not received the letter yet.

The next day was even more depressing for a couple of reasons. First, Sgt Chavez kicked me off the R-7, or M-01 as it is also known. He said Ryan and Russ had to work that day. This pissed me off because M-01 used to be my Track until I picked up rank and got a new job. Sgt Chavez was a head case anyway. He was always whining about something in his little whiny voice. We could rarely even understand what he was saying.

I am the non-commissioned officer in charge of the Asset, Tracking, Logistics, and Supply Systems version II+ for Second Assault Amphibian Battalion; in other words, I'm the NCOIC of ATLASS II+ for 2nd Tracks. This meant I had an office job; I wasn't assigned to a specific Track.

So I had to follow Sgt Chavez's orders and pick up my gear and go elsewhere. I ate supper by myself that night and then went to my tent wanting to read or write something. About halfway through my letter to Abby, we had mail call. But I never heard my name, and the rest of my letter to Abby was ruined. I just laid there motionless. I couldn't believe no one would write me. I hated everyone and I hated my life. I somehow finished the letter to Abby, but the second half of the letter wasn't good and I said some things I probably shouldn't have said.

February 25

Abby,

What's up babe? I hope you are doing okay. Things have gotten busy around here. I got my first letter yesterday and it was a Valentine's Day card from Mom. In the card she mentioned something about the other letter she had sent, so obviously the mail is not coming in order. It was postmarked on the 14th, so the date on the letter doesn't matter. Anyway, we have been working until 9 or 10 at night replacing engines and transmissions and fixing smaller things. I was able to take my first shower 2 days ago. I washed 3 times and every time I rinsed, the water still came out dirty so I finally gave up. I felt clean though, for a little while. But I just found out that my laundry is lost, so I don't have any clean clothes.

We are supposed to move out in a few days, so we'll have to see how that goes. If mail is already lost now, I can only imagine what it will be like when we're in the middle of the desert, nowhere near a camp.

Well, I thought I had something to talk or write about, but I guess not. I can't think of anything else to say. I guess I could bitch about how much I hate this place and everything that goes on within it, but that wouldn't change anything. I guess the only good thought I have to hold on to is the thought that one day, things might be better. I have a feeling that day is a long way off. I hate to say it, but even the thought of you is hard to hang on to. I guess that is because I haven't heard from you, but I can't blame you because this mail system is so fucked up. The ring you gave me is all beat up now; it's about how my life feels. I liked it better here when I could sleep all day. The time passed quicker and I didn't have to sit and think about things.

*Sorry for burdening you with my problems. The bad part
is that this is not even half of it. I hope you are doing well
and I hope to hear from you soon. Although I miss you, I'm
happy you aren't here. I love you.*

The very next day, February 26, we got more mail and I received a
letter from Dad and Abby. Finally! This was the happiest I had been
since Abby and I first met. I read Dad's first, and it was very
thoughtful and made me smile all the way through it. It's so good that
Dad and I have gotten to know each other better. I feel like I can really
connect with him now, and it's great to hear everything he says to me.

Saturday night, Feb. 8

Dear Eric,

I know you just left yesterday, but it seems a lot longer
somehow. We all got home O.K. with your mother riding
with Abby, David in the Lexus, and me driving Abby's car.
Cynthia and I both like Abby a lot. In some ways, she
reminds us of you.

Everybody called last night to find out how you were. I
was relieved when you called and we found that you were
on a big jet. Hope the flight was smooth and easy. I know
for a while you won't have too many comforts, so it's good
that it didn't start with a bad plane ride for 19 hours.

Today I washed cars. I cleaned out your truck and
washed it—it looks good. I found the new bug shield in the
garage. I assume you bought it for your truck but hadn't
had time to put in on. Tomorrow I'll put it on and replace
the muffler hangers.

Your mother is doing fine—don't worry. She is
obviously scared and sad that you are there, but she is
being a real "trooper." She thinks if you can do it then she
can too. She loves you beyond words.

In the next few weeks we will try to stay busy and not
watch too much CNN. Like you, we all have a big empty
hole inside us because of our separation from you. I know
it's hard to be away from Abby—and difficult for her to be
away from you. Just know that time will bring us together
again. I pray every day for your safety and a safe return
home.

Write us when you can and tell us about what you're doing. Also, tell Russ and Ryan that we say Hello and are thinking about them too.

I'll mail this letter on Sunday morning so you can get an idea about how long it takes to get one.

Eric, just know that I love you and my thoughts are with you always. I'm very proud of you—always have been.

Love, Daddy

Then I read Abby's letter. Her letter smells like her, and she writes like she talks so when I read her letter I could hear her speaking the words to me. It's emotionally exhilarating being able to feel her like I can from half a world away. Tears of joy streamed down my face and made it tough to read.

2-12-03

Eric —

Hey Baby! I miss you. I am really sick right now so I haven't been doing anything but sleeping and watching the news and wondering how you are. I think about you every second of every day—as always. I'm not quite sure what to write about because all the stupid things that happen during my day seem insignificant compared to the danger of war that you face daily. I am scared, not about us, but about your safety.

Okay so enough about that subject. I got my pictures back from the day you left. I sent my doubles to your mom—I thought it would make her feel a little better. She called me today to thank me for the pictures. I am going to send you a few of them so you won't forget how much we hated to see you leave. I am worried about your mom—she told me today that she hasn't left the house since you left Friday. She is worried that you will call and she won't be there. I don't think you'll be able to call! She made me feel bad—I know she didn't mean to—but I feel like I am a bad girlfriend because I don't sit by the phone and wait for you to call. I have my school, and I just started job hunting. I am planning on going everywhere until I find a new job. I can't live on nothing any more.

Everyone has been really sweet, and have called to check on both of us. I just miss you and I would rather sit at home and watch the news to see if you are okay. I get so excited when they show Marines on TV—I look at every face to see if you are there. But I haven't seen your *beautiful face yet, so if you get the chance, get your face on TV 'cause I'm looking for you.*

I am trying to figure out any way possible that I can get over there and see you. I'm thinking about coming over there for spring break. I haven't made plans yet and it's March 10-15th and there is nothing I would like better than to see you. But I don't know if it's possible because all Americans that are over there are being sent back to the US. So I thought that maybe I could get a job in photography and come over there and take pictures of the war for a newspaper or something. I'm crying—don't get mad, I just miss hearing your voice. When I think about it I always begin to cry.

Okay, so I did this compatibility thing online and I thought our results looked good so I wanted to send them to you. So I printed them out and here they are.

So they have put the USA on high alert, which means that there will likely be a terrorist attack again. They think that the east coast is the more vulnerable but I don't think so. There are helicopters flying over every 40 minutes. I'm sure you see far more than that, but they are so concerned that all flight space is being patrolled. I of course think that it will be an inside attack—like a suicide bomber or something. Every time they bomb us it has been a different way so why would they do it the same way this time. I am not afraid for my own life but I am afraid for yours.

Oh yeah—I was watching/reading something that said if we were going to war then we will probably attack sometime from 2-23 — 3-03 since there will be no moon in the sky. They said that the US has more equipment to see in the dark than the other side does. I don't know how true it is but it does make sense—and we both know things must make sense to me. So March 3rd there will be a full moon—and I plan to wish on it—it comforts me to think that you see the same moon that I see.

I love you and miss you, and I can't wait to see you.
Always, Abby

P.S. I want you to write in your journal and I want you to know that I won't read it if you don't want me to. I just want you to write about how you feel—you never show your emotions and that worries me.
P.S.S. I love you so much, blue eyes!

I couldn't wait to write her back …

Feb. 26

Abby,

You made my day! I finally got your letter that you
mailed on the 12th. The pictures are great, and you looked
cute in your little outfit. The guys here couldn't get enough
of the picture of you and me at Ruby Tuesday's. They all
say you're hot and that I'm lucky to have you. Well, they
are right, partly. I'm lucky to have you because of who
you are, not because of what you look like. Although,
being that you are hot is a definite plus.

Don't feel bad for not standing by the phone. True,
there is a phone here, but it's difficult to use. I've already
decided that I'm not going to try to call. First off, there are
limited hours of operation because it goes back to Camp
Lejeune. Secondly, if you call outside Camp Lejeune, you
must call collect which runs from $7 - $14 a minute. Next,
there is about a 5-second delay from me to you and you to
me. Also, there is a monitor that sits beside the phone and
listens to the conversation, so I don't want you to get a
false image of my mental well being. And last, the phone
when in operation will have a two to three hour wait and
also, it is down or broken more often than working. I'm
sorry if all of that disappoints you, but just know that the
reason I don't call isn't you. For the time being, we'll just
have to communicate via mail. If you happen to talk to
my mom again, you can tell her that, and I won't be
calling anyone.

On another note, I'm glad you are looking for a new
job. I hope you have luck in finding one. I'm also glad that
everyone is being sweet to you. Tell them all I said Hey
when you talk to them again. And about the Marines on
TV—well CNN was here about a week ago but it was
nothing big. I think I was actually writing you a letter when
they were here. They don't come up here too often
because we are in a highly restricted area. They have to
obtain a lot of authorization before they can come to our
position. And what's this about you wanting to come out
here? With what I have just told you, maybe you could be
convinced not to come. But then again, I know how
stubborn you are. But really, though, Kuwait is about to be
shut down. The schools are closing in just a few weeks, and
the only businesses that will be operating will be those who
support us. So no, as much as I hate it, you can't come
over.

The dates you sent me concerning our action are pretty close. Most likely it will be around the time or later than the night you will be wishing on that moon. Our stuff is to be packed tonight, for the projected march date is tomorrow. But you know as well as I do that nothing ever happens on time here, so I'm guessing three to four more days.

So I read your little compatibility test, and was amazed at how accurate it was.

I'm sorry if I scared you with that last letter I sent you. I was about to move into my "lockout" mode. This is a mode I first developed in boot camp, or at least first realized I do it. It is when I shut everyone out of my life and rely only on myself to make me happy. I guess you could call it my anti-social mode. But your letter came just in time, covered with your scent. You really wouldn't believe how happy it made me to hear from you and to smell you.

I am writing in my journal, but I'm just now getting used to it. When I began writing in it, I felt like I was writing to you.

There is so much more I want to say now, but I have to pack. I look forward to hearing from you again. I love you and I miss you so much.

Love always, Eric

P.S. The pictures you sent of you and me brought tears to my eyes. The first tears my eyes have seen since the flight over. I love you!

The next day, I wrote my brother ...

Feb. 27

David,

Hey, what's up, Man? I haven't talked to you in three weeks now. Thanks so much for coming to see me off on the 7th! It was really cool of you to do that; I know you didn't have to. I hate there was so much tension between everyone. No one really knew how to act or what to say, including myself. But that is to be expected; it's not like we can prepare ourselves for something like that. I wanted to cry most of the time, but I knew I had to be strong in front of Mom. I almost lost it on the bus, but I couldn't cry in front of Ryan and Russ. It wasn't until our flight from Frankfurt to Kuwait City that I finally broke down and cried myself to sleep.

Anyway, things out here are stupid and difficult. First off, we were scheduled to go North from here today, but that fell through. Supposedly we still have four ships that haven't come in. So it is looking like we'll be here for another two or three weeks. I don't mind that part, but they act as if we could leave at any moment and we must be ready at all times. That's what gets me, we have to pack our stuff in the morning and then unpack it at night. But once we leave here, there is no return until we meet our objectives. The final objective is a ways away from here. I write all the plans and my feelings about them in my journal. I don't feel in any way confident that they are going about this in the right way, if there is a right way at all. It just seems to me they are trying to put on a show instead of accomplishing the mission. I don't know if it would be a good idea to tell the females of our family what I just said. I try to only tell them the good things that go on here. As can be expected, they seem to be a little more weak at heart than we are. Now don't get me wrong—I'm not saying they are weak! They just don't need an excuse to worry.

On a better note, how have things been with you? We haven't spoken much lately. I've got so many questions, but have no way of finding answers. My friend Ryan got a package with food and cigarettes, and they were packed in newspaper. Believe it or not, the newspaper was the most interesting part of the package. There was something about a snowstorm on the east coast and something about a bunch of people being killed at a club in Chicago. The rest was all classified ads of Portland, Oregon. So did you get any snow? What's up with your job? Still living in Greensboro? How's Kitty? I can't wait to get out so I can get a dog. I've been reading this book that Abby gave me, Lord of the Rings, and there are a lot of cool names such as Thorin or Dain. I can't get a cat because Abby is allergic to them, unless it happened to be a tiger.

Well, Man, I'll quit boring you now. But write me back if you get a chance. Sorry I didn't write you sooner, but I was expecting to get your address from Mom and Dad. The *mail system is so messed up. I finally got some mail but it took three weeks to get here, so I haven't gotten any responses from mail that I have sent. I don't even know if it has reached anyone yet. Maybe one day, I will finally be out of this gun club and will be able to get some direction and straighten out my life.*

I hope to hear from you soon, and more importantly, I hope to see you soon. Tell everyone I say Hello. I love you and I miss you!

Love always,
Eric

On Feb. 28, we received an operations order brief. Our original plan had been to leave on the 27th, but that obviously fell through. Supposedly there were still four ships that had not arrived with our gear. So the next tentative date for leaving is March 15th.

When we receive the remainder of our equipment and gear, we will go north to the dispersal area. This will be our last point of staging before we cross over the line of departure—about fourteen kilometers from the Iraqi border.

We have five division objectives in Southern Iraq. First, air support will bomb Safwan Hill. It is the last possible outpost that Saddam has to control the border with Kuwait. As soon as it is destroyed, we will be rolling into Iraq.

Next, we have three objectives within the oil fields. Saddam has moved his heavy artillery in place so that if bombed from the air, the oil fields will blow. They've also placed gasoline barrels rigged with explosives so he has the power to detonate them himself.

Our last objective here is to take charge of a bridge that leads into the oil fields and to allow no one to pass either way.

Once all of these objectives are met, we move northwest to our first checkpoint. This checkpoint is sixty kilometers from the dispersal area and we are to reach it 24 to 36 hours after Safwan Hill is taken. From here we are to go another three hundred kilometers to Baghdad. Eight days from the beginning of the war, we are to have traveled 360 kilometers and be inside the city of Baghdad.

It seemed like the perfect plan except for a few things. Before Desert Storm, they bombed for days to "prep the battlefield." During this time, they took out key objectives and devastated the enemy into retreat or surrender.

However, this time, the first air strike is to be simultaneous with the invasion of the ground troops (us). The oil fields are to remain intact so their own income source can be used to build the new government. We cannot risk air strikes for fear of destroying the oil fields; close combat must be utilized.

The problem I had with this was the way everyone was acting like the enemy was just going to lay down weapons. They seemed hell bent on the idea that Saddam's forces were going to be so intimidated by us that they will not even put up a fight and will just show us their white flags. I was not worried about being able to beat them, but I was worried about too many people underestimating our enemy and not being fully prepared for battle.

History has proven time after time that underestimating the enemy can only lead to tragedy. Then I hear things are heating up in Korea. Apparently, this war that seems inevitable is intended to show the world what our ground forces are capable of. And honestly, I wouldn't have a problem with that if I truly believed everyone here didn't think this was just going to be a walk in the park. Oh, well, who am I to have an opinion?

My role in all of this will be with a contact team in the field repairing vehicles throughout the battle. There are three contact teams, two of which will be working and one providing security. Sleep will be limited to an occasional nap, if possible at all. The sand is very rough on the vehicles, so we have to expect the worst. We probably won't be able to get any sleep until we reach the first checkpoint.

Now that I had the intelligence brief behind me and had voiced my thoughts and opinions, it was time for me to go to sleep.

Chapter Nine

March 1, 2003. Today is my Dad's 55th birthday ... "Happy Birthday, Dad!" It was nice getting mail from him today so at least there was some sort of interaction. Plus, Dad gives me the straight scoop, and I like that.

Grandma Jones had written. The letter she wrote was hard to read but she was really sweet for writing. Even though it was three weeks old, I was so happy to at last be getting mail!

I finally got a letter from Mom. She'd sent me two cards now, and today I received her first letter. It had been written on the Sunday following our Friday departure from Lejeune. It seems like she is making every effort to stay positive. She also tries to convince herself that it's not as bad by using examples of events that happen in the States to offset the danger I'm in here. I'm beginning to wonder if she's so stressed out that she can't even think to write.

Feb. 9, 2003

To my traveling Marine:

Was hoping our days of communicating by letter ended with Parris Island! Seems kind of ironic somehow, when the internet made it so easy and so immediate—to say nothing of cell phones—that we now have neither. Especially when it would mean so much to talk with you and know what you're doing, where you are and most importantly, that you're OK.

It was wonderful getting your call Friday afternoon when you were in the airplane getting ready to leave. It's like Jimmie said ... every time we get to the point that we find it hard to breathe, something always seems to happen that lets us catch our breath—and hearing your voice for those few seconds did just that.

Eric, I really enjoyed talking with Abby on our trip back from Lejeune. She drove your truck until we got to Raleigh, but then we switched and I got in your truck with her, supposedly so she could rest from Raleigh to High Point. (I promised her I would keep quiet so she could sleep.) But she couldn't rest; we talked all the way about you. I couldn't hope to find anyone who I think could care more about you than she does. I can certainly see why you seem to feel the way you do about her.

I'm having a hard time leaving the house while you're gone. I'm so afraid you'll get a chance to call and I'll miss you! And I don't want to do anything but watch the news.

By the way, while ago they announced two Marines had been arrested somewhere up North—they had been on "unexcused absence" and were arrested for planning to bomb Camp Lejeune. Guess not all the danger is overseas.

We look so forward to hearing from you and learning how your life is going. I've never been one to feel that I could ask God for special favors, but in your case, I'm making an exception. After all, I never "deserved" much of anything, and He gave me you.

All my love,
Mom

And I got two letters from Abby! The first one was written February 8—the morning after we left Camp Lejeune. She sounded very warm, loving, and reassuring …

February 8

Hey honey,

It's Saturday morning, and I miss you. You are supposed to be laying in bed next to me with your leg bent over me but you're not here and all I can do is cry. Work last night went okay, but I kept waiting for you to show up drunk so I could drive you home—but you never came. Then when I started to drive home and realized I would have to shower and sleep alone, I started to cry yet again. I didn't think this would be so hard. All I want to do is sleep—I hate that you are so far away—I want to see your beautiful blue eyes but I can't …

… I'm glad I had the chance to ride home with your mom and talk to her. She admires the man that you are, and I love to hear stories about you. She told me about the only time in your life she got really mad at you—when you quit the baseball team. She said that when it happened she thought you went about it in the wrong way but over time had decided you had done the right thing—she admired your courage to stand up to your coach when no one else did. I love that she loves and admires you so much …

… Zula is being a brat. I think she misses you because she keeps waiting at the door for you to come home …

I love you and miss you so,
Abigail

She also updated me on some of the things happening with her girlfriends and their relationship problems—both with each other and their boyfriends.

But her letter dated February 16th wasn't all that terrific. It started off good, but then she began to scare me. In fact, this letter reminded me of the way Tammy used to treat me. Even her handwriting was beginning to look exactly like Tammy's. And now she and Brandy are thinking of going on a cruise together for spring break and partying for the weekend with all the guys in Panama City. Tammy was the Queen of "What happens here, stays here," and I can see Brandy telling Abby the same thing. With Abby, I thought I was finished dealing with things like this, but now it seems like it's happening all over again. I actually had that same gut feeling I used to get when Tammy was sneaking out and doing things behind my back.

February 16, 2003

Eric — Hey beautiful!
... This past week, sometime I got the crazy idea that I wanted to go on a cruise for spring break. It started off as just an idea, but then I started looking at prices and they were only $400 so now Brandy and I are going on a cruise for spring break. I couldn't decide if I should go or not but I know I would regret not going anywhere for my last SB. All of the guys are going to Panama City but I don't want to go there—I want to go somewhere I can relax. I invited my mom, my dad and sister to all go because I feel like I need as much support as I can get right now but they aren't going to go. So it's just me and Brandy! I wish you could go. I have always wanted to go on a cruise and I'm excited that I'm going—I just feel like you should be able to share in this experience with me. I don't like deciding on major events in my life without consulting you first. I know I should be able to do it on my own; I just wish I could ask you for your opinion. The main reason I decided to go was that if you were here at Camp Lejeune you wouldn't be able to go with me anyhow.
We leave Miami on Monday, March 10 and arrive back in Miami the following Friday. Brandy and I are driving down to Miami in my car—let's hope it makes it. We are planning on leaving

Charlotte on Friday afternoon, March 7; Brandy wants to stop at Panama City and party with the guys for a few days. Then on Sunday, the 9th, we are driving down to Boca to stay at Troy's parents' house for the night so we won't have to drive so much on Monday. I hope we have a good time. I just wish you could go with us ...

Always & forever,
Abigail

I was wishing she wouldn't go, but couldn't really tell her not to. If she goes, then it will be hard for me to trust her. If, on the other hand, I tell her I don't want her to go, she'll blame me if she regrets not going. I will be the reason she "missed out" on her last UNCC spring break and she will use it as ammunition against me to all her friends, including the guys that would like to be more than just a friend to her. Besides, I don't want her to think of me as being overprotective or insecure.

When I got back to the states, I was hoping to take her on a cruise with the money I will be able to save while I'm here. But I guess that's out of the question now. Maybe Ryan, Russ and I can go on a cruise together and see how she feels about that.

I couldn't even tell her if I wanted to. She leaves on March 7th to go to Panama City so if I were to write her a letter and describe what I'm feeling right now, she wouldn't even get the letter until after she was already back from the cruise! I'm pissed now, and helpless to do anything about it.

"Fuck it, I'm going to sleep!"

The next day, March 2, I woke up and had the bright idea to make a postcard to send my Mom and Dad. I made it from a piece of the cardboard that is used in the cases of water that we drink here in Kuwait. The Arabic label was printed on one side; the other plain brown unprinted side made me think of the way our surroundings look and I thought of it as a picture postcard.

"I hope that you enjoy this field expedient postcard that I was able to obtain from one of the many boxes of bottled water that we receive daily. The postcard, however very plain, is actually a better representation than this place really is. Simply

smear sand, dust, and sweat to get a better depiction. I hope to see you soon, and love always, Eric."

ERic Cox
2D AABN, H&S Co, Det D, Camp Matilda
Unit 76689
FPO AE 09509-6689

FREE

CARD TRANSLATION:

OASIS

PURE DRINKING WATER

JIMMIE AND CYNTHIA COX
1107 N ROTARY DR
HIGH POINT, NC 27262

I HOPE THAT YOU ENJOY THIS FIELD EXPEDIENT POSTCARD THAT I WAS ABLE TO OBTAIN FROM ONE OF THE MANY BOXES OF BOTTLED WATER THAT WE RECIEVE DAILY. THE POSTCARD, HOWEVER VERY PLAIN, IS ALTUALLY A BETTER REPRESENTATION THAN THIS PLACE REALLY IS. Simply smear sand, dust and sweat to get a better DEPICTION. I HOPE TO SEE YOU SOON, AND LOVE ALWAYS,   ERIC.

A couple of days passed. I tried to think about things other than Abby's plans for spring break.

March 5

Mom and Dad,

Hey what's up? So not much has changed here—only dates. We hear a lot of aircraft flying north over us at night, and returning shortly thereafter. We heard on Kuwait radio that Saddam had a military parade in Baghdad today and said he would win any war. We could only wish that we were in attendance at his parade.

I got a few of your letters today. Thanks for sending me the race results. Haven't seen McGrath competing? You may find it interesting if you don't already know it that Jeremy retired last year; his last race was in Europe. But the weird thing, besides giving up his $750,000/yr. 4-year contract with KTM, is that he's spending this entire year preparing for the 2004 PGA Tour. He reportedly even stated that he could beat Tiger Woods. However, he is still working with KTM in product design and engineering, and rumor has it that KTM is coming out with a golf cart. He is also talking with Bud Light for sponsorship in golf, saying that there are just as many beer drinkers in golf as there are in motocross. On another note, 125 East Supercross winner Branden Jessemen must be doing well for himself. Remember him? He and Ben Riddle battled together in the 125-B class at the Arenacross in Winston-Salem. Can you tell I'm still a fan?

Well, the 60 days of leave that I had planned on selling back to the Marines can't be done unless I reenlist or get out in less than a month. So I'm going to have about 80 days of leave when I come home. I have to take 30 days before October because by then, I'll have 90 total days, and anything over 60 will be cut off at the beginning of the new fiscal year. But with my luck, they won't let me take it.

And yeah, the bug shield is for my truck. I just kept putting off putting it on. Usually the only time I thought about it was when I was in Charlotte, and it was at Lejeune. Would you believe that I bought it back in June? Yeah, probably so ...

I miss you all and hope to see you very soon. When all this kicks off, you might see me on TV working in the petroleum industry. But look fast—we won't stay there for *long; there are more important missions at hand. Keep in touch. I love you!*

Love always, Eric

And then I wrote Abby.

March 5, 2003

Abby,

What's up, hunny? Not much has changed here, only dates. They keep pushing us back further. Sorry I haven't written in a few days. We have had some extra duties because we got in trouble the other night. We were having wrestling matches and it got out of hand. Master Guns heard us from his tent and came over to ours and put a stop to things.

I got two of your letters the other day. Don't worry about not knowing what to talk about. The more you write, the better. You've got to know that receiving mail over here is the best part of the day. I read your letters over and over. I've already read all of the magazines that are here, and I also finished the first part of Lord of the Rings.

I'm sorry things didn't work out with you and Nicole, although I didn't like her anyway. I know you say that she was the sister you never had, but she's not. She's nothing like you. Anyone who could do what she did to you, is of no relation to you, and not worthy of having you as a friend. Patty is a good friend to you—although I don't approve of her ways—and Kate is a good friend to you, although you may not choose to want her. Good friends are hard for a girl like you to find. I say this because close peer groups are usually connected by the ability to relate to one another or having similar personality traits, likes/dislikes, and beliefs. And you are by no means your average, ordinary, run-of-the-mill girl. The caring, love, and compassion that lies within your heart is something that other people will know of only by reading romance novels.

That's cool that you are going on a cruise for spring break. I bet you can't wait! I hope that the weather is good for you. It's kind of funny that you decided to go on a cruise. It turns out that I can't sell my leave back over here like I had planned. So I'm going to come home with about 80 days of paid vacation. So of course, I have been trying to decide where I would take you. The funny thing is, I was planning on taking you on a Caribbean cruise! I thought it would be nice to go and experience something new together. Well, by the time you get this, you'll have already gone, so you better write me and tell me all about it. I at least want to experience it through your letter.

So how is school going? Are you still on track for graduation in May? Are you going to the real estate school? I wish that I could be there to watch you walk across that stage! Have you sent an invitation to my parents yet? I'm sure they would love to go. What are your plans afterward? Still moving to Wilmington? I've decided that I will base my decision on which college to attend on the location you decide upon. ...

Well, I'm going to eat now. Tell everyone that I'm doing fine out here and that I say, "Hello." Oh, who are the guys that are going to Panama City? Aaron, Patrick, Richard? If Aaron, are he and Jess broke up again? He seems to always find a way of breaking up right before spring break. And who's Troy, and where's Boca? Anyway, it's cool of his parents to let you stay. I miss you so much! I can't wait for the day when you're back in my arms again. I love you so much. Promise me that you'll be careful over spring break—I don't know what I would do if something were to happen to you.

Loving you,
Eric

Then I couldn't believe it when the evening before Abby was to leave for Florida, I had an opportunity to make some phone calls home. I talked to my Dad, my Mom, and thank goodness, reached Abby! I couldn't wait to write her afterwards.

March 8

Abby,
... I'm so happy that I was able to talk to you Thursday night! They told us they had fixed the phones to where they worked a lot better, so I gave them a try. Unfortunately, they didn't work any better—they frustrated the hell out of me. But once I was committed, I wasn't going to quit until I succeeded.

I got there at about 10:30 that night—2:30 PM your time. I tried my parents first to ask them if they had sent you the flowers I had asked them to send. It took about 15 minutes to get through, and then, I got their answering machine. At this point, I had 5 people in line behind me, so I moved to the back of the line. At about 11:30, it was my turn again. I tried home once more and got connected with Mom. She was so excited and relieved to talk to me. She told me that she was going to shut up and all she wanted to do was to hear how I am doing. So I told her everything I could think of for about 5 minutes and then we got cut off. At this point, I had 3 people behind me so I went to the back of the line again.

At about 2:00 AM, I got back on the phone. I tried to use the credit card to call your cell. But just as it began to connect, it cut off and said the credit card did not allow calls to be made on it. So I tried to call your apartment but no one answered; still it made me feel better that I heard your voice on the answering machine.

So then I attempted to call Mom to see if the Visa card number had changed. Dad answered this time; Mom had gone to the grocery store. I had a good talk with him. He said the credit card number was the same, so my hope for reaching you on your cell phone was gone. Dad asked me if I wanted him to give any messages to anyone, so I told him to tell you everything I had said and that I love you and I'll be home soon. Then it hit me. I gave him your number and told him to call you on your cell as soon as we hung up and get your roommate off the internet because I was trying to call. He did, and it worked!

I was scared at first because I didn't know what I would say or how I would act. But my worries were gone the instant I heard your voice; you seem to have a way of making my life perfect. I must admit, I was terrified of you going on spring break. I couldn't get it out of my head, and there was nothing I could say or do about it. A few days passed before I finally got over it, as I always do, and decided to write you. Then I had a chance to call you, and did.

When I came back to the tent, I laid down and Ryan asked me if I got through. I was so happy he was awake (or that I had woke him). I told him all about it. I knew he didn't want to hear all I told him because he kept falling *asleep. But he did ask me if I was still worried about you. I said, "Worried? When was I ever?" ...*

It's funny how things work out, and how you seem to always change me for the better. On the other hand, it scares me to think that someone has so much control over my life, without controlling it. You make me happier than anyone in the world. Since my goal is happiness, and you bring me happiness, it's safe to say my goal in life is you ...

Right now, it's 10 AM Saturday morning in Panama City. I hope you had fun last night and hope you will have fun for the remainder of the weekend and fun on the cruise. I love you so much and cannot wait to come home to you.

Love always,
Eric

On March 10, I got a care package from Mom. One of the items was eye drops, but they were the wrong kind because I was still wearing my contacts. I wrote her to thank her and explained ...

" ... Thanks for the care package. I had no idea that you endorse tobacco use. Ha ha—I'm just kidding—they help pass the time when it's really boring. All of that stuff will be a great help to me. The only thing is the eye drops. They cannot be medicated (red eye relief); they must be re-wetting drops. Yes, I'm still wearing the same contacts— don't really remember when I put them in. Probably late December or early January, but they don't bother me so I'm good. I don't want to wear glasses because we have so many gas attack scares, and when I wear a gas mask, I can't see without my glasses."

And I told her I was pleased that she had talked with Sgt Pit's Mom when his mom contacted her to pass along some information she had received about our unit ...

"Anyway, you met Sgt Pit's mom, Cherie´? That's funny—he's in my platoon and one of the very few guys that I hang out with. He says his mom is more motivated than he is. Kinda reminds me of you. Keep in contact with

her—she's done this a lot and has many connections for finding out info. Pitkovich ("Pit") says Hi, and he apologizes if his mom is annoying you. I assured him that she wasn't."

I wondered if his mom knew how much her contacts meant to my Mom.

I also told her about the problems Abby and I were having with letters being delivered in no particular date sequence. The most recent ones were often being delivered before those written earlier. It was so frustrating when it made it look like a question or issue, or even a fear raised by one or the other of us was being ignored.

And finally, since Tammy had recently told me to tell Mom hello, I told her about the phone conversations between Tammy and me just days before I left and about the letter I had written her at Abby's suggestion when I got to Kuwait to explain my new relationship. In my letter, I had reiterated the fact that I was Abby's, and wished her luck with her friend/boyfriend, Luke, saying that I hoped she had found as good a person in him as I had found in Abby.

On March 11, I received a letter in a care package that Abby had sent me just days after we left the States. At the time the letter was written, she had been a little upset that I had not called her when she later learned Russ had had an opportunity to call home to his family. It seemed as though she was somehow questioning me, and wasn't concerned about the hours of effort it normally took to make just one three-minute phone call. I wrote back to explain the circumstances. And I knew as I was writing that she was in the Caribbean.

" ... On another note, it's Tuesday afternoon where you are. So how's the weather in the Caribbean? It's hard at times for me to comprehend the fact that you are laying out in the sun having a few Martinis, floating on a cruise liner over tropical waters while I lay writing you from a tent in the middle of a desert, just miles away from a country full of men with guns pointed at me. So when at times you think you have it bad—just think of me. And at times like these, when I think I have it bad, I just think of the boys that are about to be killed by my comrades and me. I guess what I'm trying to say is that it could always be worse. Just know that when I get home or shortly thereafter, I'm taking a vacation—I'm overdue ..."

I also thanked her for the cookies she sent, and for the inflatable pillow. The pillow had been sort of a puzzle at first; when I wasn't sure what it was and asked several others what they thought it was I got answers including a portable shower, stadium seat, a water dispenser, and flotation device. But when I decided to try it as a pillow, I found I slept much better because of it.

On March 16, I got two letters from Abby. She was continuing to have serious relationship problems with her mother; I tried to advise her as best I could. But trying to help her deal with her problems made me think about my own.

"... There, you have my thoughts. Speaking of thoughts, you asked if I have trouble sleeping at night. You seem to read deeper into my letters than I thought was possible. If you must know, yes, I do. I wish you wouldn't worry about me so much. But since I know it would make you happy, I will give you a small analysis of myself.

First of all, over time I have trained myself to deal with many things. I know what I must do to deal with problems, manage my anger, and rid myself of stress. I have the knowledge to and will survive. With all that being said, during trying times, I may for a brief moment prove to be helpless. In reality, I hate this place and desperately want to go home. Not only do I hate this place, I hate everything about it. I hate my job, the environment, living conditions—everything. I hate that I'm not where I want to be and have no control over it. I hate that I'm not with you and all the other people that I love in my life. And I hate that I have to bring you and them into all of this and make excuses and write letters that are emotionally targeted to justify my being here and to ease their fears.

But when and only when I begin to bring these pessimistic ideas into my daily routine is when I deteriorate mentally. So being the optimist that I am, I find good in everything I do. Sometimes it is very difficult to do, but I always manage to do it.

As for the difficulty of sleep at night, I don't mean that it is hard to go to sleep. I fall asleep easily as my body is trained and accustomed to do. But when I awake from a bad dream in complete darkness, restricted in a sleeping bag and small 2-man tent and not a clue of where I am, I become claustrophobic and freak out trying to escape. Luckily, once upon my feet, I usually come to my senses, kiss my ring (you) for comfort, and go back to sleep. That is one instance of when I may prove to be helpless. It happens rarely, though, because I'm not fully conscious ..."

Having had enough of self-analysis, I wanted to talk about more current issues. Abby had mentioned another contact with Russ's girlfriend, Kate, when Kate had visited and shown her some pictures she had taken of Russ, Ryan and me at Camp Lejeune the week before we left.

"Enough about that! How is Charlotte? So you talked to Aaron and Jess? I got his letter and he couldn't say enough about you. He said he and Jess were trying to cheer you up. That's really sweet, and I'm glad that you finally became friends. But Baby, Cheer Up! I'll be coming home to you soon, I promise!

And what's this about Kate? How was supper? Does she not like Russ? Ryan and I spend a lot of time analyzing her, and we don't come up with anything good. First of all, we think she has ulterior motives. Why did she hook up with Russ in the first place? Was she trying to make me jealous by hooking up with my friend? Did she go to Camp Lejeune to make me jealous or to be in my vicinity? Does she send him nice things to make me jealous while writing him dry letters? Is she in contact with you because of me? Is she befriending you to try to cause drama in our relationship? Why did she decide to join the Marine Corps? She now says it is something she has always wanted to do, but in the entire time she and I were together never even mentioned it.

The bottom line is that Russ fell head over heels in love with her in a matter of weeks, but now questions her love for him. If she is toying with him in order to stay linked to me, she will be making a huge mistake. I try to believe that none of my thoughts are true, but even more, I try not to let Russ believe them either. It's difficult though. If he and I are correct with our assumptions he should cut all ties with her. But at the same time, we could be wrong. Maybe we're just falling into the throes of this forsaken place and chancing throwing a meaningful relationship out the window without any confirmation from her.

Anyway, she's already in boot camp so it doesn't matter. I think that fact alone will cause them to break up. Tell me your thoughts on this. I wrote you earlier I thought she was a good friend to you, but now I question that.

... Okay, so you win the award for getting the longest letter I've ever written. Go fast as you read. Please stop worrying so much about me—I'll be fine, and I'm still the same ole' Eric that holds the key to your heart as well as your dog's collar. Ha ha ... tell Zula I say "Well Hello," and that I miss her. I love you, more.

Love always, Eric

March 17. Today was a slow day of sitting around and doing nothing as we staged our vehicles and gear and prepared for departure. I have to ride on a 7-ton truck, which is more comfortable, but not as much fun as the R-7 Track that I'm used to. The AAVs are fun to drive because they're almost indestructible and it takes a lot to get them stuck.

I rode four-wheelers and dirt bikes all my life. I got my first four-wheeler when I was four years old and my first dirt bike when I was five. I began racing competitively when I was fifteen.

Looking back now, the quick three and a half years that I raced were the best years of my life. Racing is a sport like no other; it's all about pushing limits. When I'm racing, nothing in my past or future exists. All that matters is that one instant.

I close my eyes and envision the start of a race, and my heart begins to pound. I see the girl holding the thirty-second board turn it sideways to signal two to eight seconds until the start, and then she runs off the track. Throttles twist wide, engines rev to the max, all forty gates fall, clutches drop, and it's fist full of throttle, wide open, handle bar to handle bar, shifting through the gears, and diving into the first turn. Being the first one into the turn means you have taken the "Hole shot," and for me there's no better feeling in the world!

I don't question my decisions and I have no regrets in my life. No, wait, I take that back ... often I wonder why I do what I do but I never regret anything I've done because I know that for whatever reason, I was meant to have made that decision. I have to tell myself these things.

My friend Nik makes a note in my journal ...

Hey Eric,

What's up? Have a good WAR

Love
NiK

We received a letter from the Commanding General prior to leaving for the line of departure. It was somewhat motivating, but at the same time, lacking a little. When I thought back to what I knew of motivational efforts from officers and Generals to their troops heading into war, I didn't remember anyone writing a letter instead of giving an actual speech. Maybe some did, though; I could be wrong.

March 2003

1st Marine Division (REIN)

Commanding General's Message to All Hands

For decades, Saddam Hussein has tortured, imprisoned, raped and murdered the Iraqi people; invaded neighboring countries without provocation; and threatened the world with weapons of mass destruction. The time has come to end his reign of terror. On your young shoulders rest the hopes of mankind.

When I give you the word, together we will cross the Line of Departure, close with those forces that choose to fight, and destroy them. Our fight is not with the Iraqi people, nor is it with members of the Iraqi army who choose to surrender. While we will move swiftly and aggressively against those who resist, we will treat all others with decency, demonstrating chivalry and soldierly compassion for people who have endured a lifetime under Saddam's oppression.

Chemical attack, treachery, and use of the innocent as human shields can be expected, as can other unethical tactics. Take it all in stride. Be the hunter, not the hunted: never allow your unit to be caught with its guard down. Use good judgement and act in best interests of our Nation.

You are part of the world's most feared and trusted force. Engage your brain before you engage your weapon. Share your courage with each other as we enter the uncertain terrain north of the Line of Departure. Keep faith in your comrades on your left and right and Marine Air overhead. Fight with a happy heart and strong spirit.

For the mission's sake, our country's sake, and the sake of the men who carried the Division's colors in past battles-*who fought for life and never lost their nerve*-carry out your mission and *keep your honor clean*. Demonstrate to the world there is "No Better Friend, No Worse Enemy" than a U.S. Marine.

J.N. Mattis
J.N. Mattis
Major General, U.S. Marines
Commanding

TWO

Chapter Ten

"Goin' to country!" This little saying was big with Ryan, Russ and me with anything we were doing. We tried to make it fun and exciting by being enthusiastic. It meant we were going to Iraq.

On March 18 at 0230 local, an order came down for 2d Tracks to vacate Camp Matilda and head northwest a couple of clicks to Logistic Staging Area-5 or LSA-5.

0430: The quartering party is woken up and told to pack their gear—they are leaving.

0445: Everyone is woken up and told to pack their gear and be at the maintenance tent at 0530 to have ammo issued.

0600: With 290 rounds of ammunition in possession, Ryan and I leave to go eat breakfast.

0700: We have a Company formation and then break off to our respective vehicles for departure. Incoming and outgoing mail has ceased for the time being.

Finally, at 1500, we left out as one large convoy to LSA-5 at such a slow pace you could walk faster. Literally, we were moving at two miles per hour for long periods of time.

At 1535, we drove past LSA-5 but no one was there. Confused, we wondered if maybe that wasn't the correct location and continued moving. But LSA-5 is only five clicks, or about four miles, from Camp Matilda and it's already been more than thirty minutes since we began on our convoy. The dispersal area is fifteen clicks away from Camp Matilda.

Fifteen clicks into our journey, LSA-5 was without a doubt behind us. At 1830, it was completely dark and we had traveled about twenty-five clicks. Everyone thought we were lost, and they were absolutely right. Luckily, we strayed only a few clicks from our path and were able to make the dispersal area by 2330.

Again, we were the only ones there but began to set up security at 25%, which meant that 9 men were on watch as 27 slept. We had to sleep together as a group and dug holes to sleep in so in case a Track ran over us, we could hopefully survive.

I slept for about an hour before I had to get up for my security watch. Then I got into a quarrel with the Molester, as he had woken me too early for my shift. With only an hour of sleep I was extremely irritable, plus I wasn't very fond of Cpl Lester.

Lester, Lester the Child Molester was another one-up kid. If you had gone to the moon, well, he has had coffee on Mars. He got his nickname from the fact that the only girls he ever went out with were the ones still in high school.

March 19. Although sleep deprived from the one hour of sleep the night before and four hours the night before that, Wednesday would go much better. I learned Sgt Chavez had been left behind at Camp Matilda. So I spent the morning on the R-7 with Russ and SSgt Little. Then we went to another position to repair some AAVs. It was nice getting away from everyone in our platoon for a little while.

Upon our return, we found our Battalion had moved north a couple of miles. We rejoined them, set up security, and once again dug sleeping holes. Grimsley began to lose it mentally, saying we were digging our own graves. But he managed to get over it rather quickly when most of the guys laughed at him.

I got lucky and had the first watch shift from 1630 to 1900. Then I slept until 0130, when we were woken up to a possible Scud attack. We got into MOPP Level 4 for a short time until the attack was cleared and we were out of danger.

MOPP is the acronym for Mission Oriented Protective Posture. What it really means is that MOPP gear is our barrier against Nuclear, Biological, and Chemical (NBC) warfare. And really this suit only protects against the most common chemical agents, which are nerve, blister, blood, and choking. MOPP Level 4 is the highest level. Gas mask is on and no skin is exposed.

At times like these, I wonder what would happen if a Scud missile really landed. We can't get up and shoot at anyone because there would be no one around to shoot at. All we can do is lay here and pray it doesn't happen. If it does happen, all I can do is pray that it hits me directly and that I die immediately without suffering.

I don't remember ever being afraid of death or dying. I truly believe that through good works and faith in God, I will be blessed with whatever is in store for me after I die. However, I am very afraid of the method in which it could take place. I couldn't imagine drowning or being burned alive. I couldn't imagine being tortured, or losing limbs and slowly bleeding to death. I want to die in my sleep at my most peaceful state. If it has to happen anytime other than in my sleep, I wouldn't mind as long as it's instantaneous. Again, I know where I'm going; I just hope to stay here in this world until I have time to do most of the things I want to do.

March 20. We switched over to Zulu time the other day, so it's now 0415 instead of 0715 and the sun is up. I crawled under a 7-ton to find some shade and tried to catch up on some sleep. I woke to find several people sitting near the truck talking and riddling, so I joined them.

It was nice to talk and laugh with people other than Ryan and Russ. Along with the jokes and riddles, we told stories about some of those we were with.

It was even funny talking about "taking a shit." (It may seem to be an indelicate subject, but when you're in the position we were in it becomes a very important part of your life.) Our Navy Corpsmen have a box for us to sit on. They pick a spot downwind from our position and dig a hole just smaller than the box to sit over. The rectangular box has four sides, no base, and a lid with two holes in it for you to position yourself on. The two holes are there so two people can sit at the same time.

It's just shy of hilarious to see two guys sitting on a box back-to-back with their pants around their ankles—well, at least it's funny until it's your turn to have to go. Once you're there you can face one of two ways ... away from everyone with your naked ass facing them or towards them and being able to see them watching and laughing at you.

When you figure out which position you're more comfortable with, in come the flies. The flies around here have a keen sense of smell for this sort of thing and flock to the box in record time. They fly down in the hole and frolic around until someone comes and sits on the box, sealing them in. Then the flies land on their butt and crawl around on them while they try to eliminate. So the person slides forward on the box in an attempt to give the flies an escape route. But the flies are happy with where they are. Even the simple act of wiping one's self becomes an issue; wiping has to be done gently and very carefully to avoid squishing a fly where one does not want a fly squished.

But though we laughed about it, it was really quite degrading. Flies that had been feasting on all our human wastes were visiting us where we were most vulnerable. People were watching and laughing. I thought back to episodes of Charlie Brown where the dirty kid always had flies around him. I thought of myself in this same way and felt that other people were viewing me in the same manner.

After my first time, I waited until dark to take care of my business because the fly population was slim. Also, I couldn't see anyone looking at me, although I knew that with the night vision goggles (NVGs), they probably were. The downside to this was the fact that nightfall usually meant sleep time. And going to the bathroom sometimes took away quite a bit of sleep time because there was usually a line to wait in. I trained myself to only have to go about once every three or four days.

That afternoon we watched as our artillery was aimed north towards Iraq. There were six rounds of what looked to be rocket-assisted artillery that went off just northwest of our position and flew slightly northeast traveling over the horizon.

At 1615, we had a formation to receive any final word. The word was that at 0020, the first serial would be staged; at 0030, the second serial, and at 0040 the third.

At 1700, I had to man the 240G in the fighting hole and then minutes later, lights began flashing over the horizon. There were rockets flying way above our heads that went from the southern horizon to the northern horizon. We could see and hear all types of explosions and I began to count the elapsed time after flash to see how close we were. There were so many flashes and booms that it was difficult to calculate because I didn't know which boom went with which flash, but from the best I could tell, we were about six miles south of the action.

We slept for a short time, but were woken up at 2300 to prep the truck for staging. We were to roll out at 0130, but at about 0010 that time was postponed to 0330. At 0345 we rolled out and traveled for about a mile but then stopped for an extended period of time for an unknown reason ... just as the convoys had done since we began. Five hours into our journey, we had traveled 1.7 miles.

Chapter Eleven

At 1100 on the 21st of March, we entered Breach Point 3 of the berm. There was a fence about 8 ft. tall first, then about 15 ft. further there was another fence about 12 ft. tall with a triple strand of concertina wire at the base, a double strand halfway up the fence and a single strand at the very top. There was also a 9-cable electric fence encasing that fence. And about 15 ft. further, there was another 8 ft. fence.

We crossed the border into southern Iraq. Oyster, Scheider and I sat in complete silence, scanning the horizon for movement. We fully expected to be ambushed or attacked as we invaded their country. Up on the hill ahead, we could see some sort of military equipment that looked staged for defense. It was smoking so we guessed it had been blown up last night in the light show. But when we got up close to it, we could see it was rusty. Apparently we were in a graveyard of fallen vehicles left over from Desert Storm. It must have spooked our assault units as they drove through just as it had spooked us.

This was right about where the traffic was jammed as Breach Point 1 bottlenecked with Breach Point 3. There were about five convoys besides ours, all trying to get through. We waited for the first serial and the quartering party as they were tied up in traffic from BP1. As we waited for them, a LVS and a HMMWV broke down so we had to repair. We waited there for about an hour before we were ready to move again. There was a lot of tension among the troops as we had barely crossed into enemy territory and already our vehicles were breaking down.

Off in the distance we could see the fires from burning oil fields. Looking toward them, no one knew what to expect.

Suddenly, out of nowhere just as we were about to roll out, a 40mm grenade round landed in front of our lead vehicle just three vehicles in front of us. I went for my M16 and scanned the area for possible enemy. I could see none. Our radios were going crazy. The second vehicle signaled to mask up. The round did not explode, so they assumed we were under chemical attack. Everyone held their breath, put their gas mask on, cleared the bad air from their mask, and gasped for a new breath.

I grabbed and displayed the orange vehicle flag that signifies an NBC attack. Transmissions over the radios were difficult to understand with all the chaos. No one knew where the shot came from, and that was the most terrifying part. We were just sitting ducks not knowing which way to aim.

"Get us out of here!" one Marine yelled.

"Go to the defense!" another screamed.

"Man your positions!" someone confirmed.

"Shots fired at my vehicle!" a Marine from the second vehicle exclaimed.

Other transmissions were broken and unreadable. No one knew what to do. We were taught to either drive through an ambush or to get out and set up a defense. This was all useless now because no one knew what situation this was and therefore didn't know how to react to it.

Then just seconds after this all started but after what felt like hours, CWO2 Riggs transmitted, "All clear, unmask."

CWO2 Riggs is our Battalion NBC officer but I wondered to myself, then out loud, "How can someone from a mile behind me say that everything is O.K.? I mean, I saw the round hit in front of me!"

"Hey, everybody, listen up!" MSgt Landon, our Company Master Sergeant, announced in his own persuasive manner. "There was a negligent discharge on one of the Tracks. It's all clear, unmask!" With that being said, everyone calmed down. Then we all laughed and joked about what just happened and the looks that had been on each others' faces.

MSgt Landon was here eleven years ago in Desert Storm, and he's one of those people that when he speaks, people listen. He's either got something really funny to say or something really important to say.

He's an average size black man that speaks with a deep Southern accent, walks the pimp way, always smokes Newport Full Flavors, and the phrase "God Damn" has a leading role in his vocabulary. You can't help but want to be around him because he's just that much fun to be with and has such an influence. Not that everyone wants to be a swearing southern pimp smoking Ports, but it just adds to his act and is very entertaining.

Although this freak accident actually helped us make light of the situation, it delayed us for some time. The glow from the fires grew brighter as daylight was slipping away. We made our way toward them as it turned completely dark.

The fires were so enormous they had given us a false sense of distance. We were moving at a steady rate of speed and it seemed like hours before we got close. Within a mile or so of them, I felt warm black specks of oil falling on my arm hanging out the window. Nearing closer yet still at least a thousand yards away, I could feel the blazing heat as if I was standing next to a bonfire. From hundreds of yards away, these colossal fires were so hot that it actually burned our faces and eyes to look directly at them.

The nights in the desert in March can get very cold. After passing and moving away from the fires and their heat, there was a drastic drop in temperature. It became freezing cold. The smoke from the fires left the night pitch black. The blackout markers and interior gauge lights were off. And to make matters worse, the vehicles in the convoy were producing clouds of dust. Sight distance with NVGs was limited to about 10 ft. at times as we tried to follow the vehicle ahead of us.

On a strange encounter, passing us in the opposite direction was a herd of camels. I counted about eighteen that were full grown, two babies trying to keep up, and two massive elders striding to bring up the rear. They were only ten ft. or so from our convoy so I tried to get their attention by speaking to them. We jokingly assumed they had seen enough up north and were migrating south for safety.

Shortly thereafter, Scheider was getting drowsy so I took over driving for him. I had never driven a 7-ton rig, but I thought to myself that learning to drive this monster in almost zero visibility was as good a condition as any. It was actually quite fun, although I almost slammed into the vehicle in front of me several times.

A couple of hours later, we finally stopped to set up our first BSA. I think that stands for Battle Staging Area but no one around us really knew … it was just the best explanation we could come up with. There are so many abbreviations in the military that I don't think any one person fully understands the entire system.

I was tasked with setting up security for our BSA, which was a simple task that I completed immediately. Two hours later, the entire BSA set-up was complete, and I tried to catch a few hours of sleep under the truck. I woke up shivering; the wind had really picked up. I tried to take the smallest possible form in order to retain some warmth but it just wasn't happening. Miserable, I climbed into the cab of the truck with Scheider, and we both slept in the upright position for another hour or so.

March 22. We were beginning the day by shaving and brushing our teeth with the small amount of water we had when a family of Displaced Citizens came within a hundred yards of our position. There was a man, woman, and three children. Their unexpected appearance caused a few of our troops to get into trouble and ended up being quite entertaining. I have to admit that I was a little uneasy, too. No one knew what to expect and we had been taught to trust no one in the country.

Talibe, our Iraqi translator, a few Staff NCOs and about 8 troops for security went to question the DCs and keep them from getting too close. What intelligence was gathered, if any, I do not know.

But what was happening simultaneously was hilarious. A SSgt from Supply was trying to take charge and had gathered a fire team of 6 troops. They began rushing these DCs while our translator was already talking to them with his own security.

1st Sgt Short flipped out and began yelling. It's always funny to watch him yell because he does it in such a heartfelt manner. But he was further away from them than I was, so I helped out by relaying the message for them to stop what they were doing and come back. Velasquez and Hayes were heading back, so I assumed that everyone else also heard and would follow suit. I turned around and continued shaving.

"Get the fuck back! I've done seen it all now!" screamed 1st Sgt Short. "This company's a bunch of clowns!"

Apparently the others from Supply had not listened to my cue, and now 1st Sgt Short was on his way to meet them. He was cursing them in the only way he knew—non-stop. He passed by me on his way to them, cursing to himself as he went. This made the lashing even worse because everything he said to himself just pissed him off even further. When he finished verbally assaulting his guys he walked back to his vehicle, still cursing and shaking his head in disbelief and I'm pretty sure it didn't end there. But that's just his personality, and it's hilarious to watch if you're not the target of his anger.

After all the excitement, we departed our position and made our way to BSA-2. I slept about an hour and a half, which was most of the way. We set everything up, and with the daylight still remaining, were to repair and perform maintenance on the vehicles.

Ready to begin working, I spoke to Ryan briefly about the convoy there and what we had experienced.

"Cox, you need to be turning some fucking wrenches!" Lope Dog yelled at me.

So I did as I was told and then he yelled at me again. "Cox!" he shouted, "No one touches any vehicle without first clearing it with SSgt Benson!"

This is the kind of stuff that drives me crazy so I did nothing for the remainder of the day.

At nightfall, I went to sleep for a couple of hours under the truck with Nik. We both had security watch together that night and slept nearby so we could be found.

March 23. We missed the mark for departure by 15 minutes. We were heading northward toward the town of An Nasiriyah. After rolling for about 6 hours, we passed a group of heavily guarded Enemy Prisoners of War (EPWs). About an hour after that, we pulled off to the side of the road. We sat for about an hour while I wrote in my journal and tried to squeeze in a nap.

Just as I got comfortable, we started up again. As we turned off the highway, we had to stop on the exit ramp for two herds of sheep and ten shepherds herding them across. Once more underway, we realized we had made a wrong turn and had to make a U-turn in the middle of the highway, really making us look like idiots. Heading in the right direction this time, we passed more DCs with many children waving and cheering us on which we found touching.

Not far from there, we pulled over again as there were two other convoys pulled off to each side of the highway. Supposedly someone had spotted movement over the berm to the east and they were assessing the area to deem safe for us to pass.

At the same time, Regimental Combat Team-5 (or RCT-5) was taking sniper fire a couple of miles ahead. We were supporting RCT-5, so we could not move ahead of them or for that matter, even up to them.

I couldn't believe all this was happening just up the road from us. I heard the shots and saw the explosions, but was not allowed to advance to do anything about it. Here we sat waiting for the area to be secured. Then we got word that we'd be staying for the night and leaving at 0400.

March 24. We were up at 0300 and ready to go by 0330, but at 0345 we got word from 1st Sgt Short that we wouldn't be leaving until mid-day.

At 0430, we heard seven were killed last night and there had been fifty casualties. Three Marine prisoners of war and more Army POWs were taken. My initial fears and doubts that the Iraqis would just let us walk through their back yards with no resistance had apparently been well founded. And after this news, we learned the reason we weren't leaving until mid-day was because they were still fighting. It was going to be a long day with few words spoken. Troops began acting irrationally.

At 1515, as I was studying a book on Iraq, I heard two shots fire and then two explosions nearby. I looked up to see Marines running around in the alert position so I reached for my rifle and ammo as I dove out of the HMMWV to defend the west flank. I could see nothing threatening, but then I heard shots ring out from the 50-Cal Machine Gun just thirty yards to my right.

There was a house about a hundred yards away in the field that we were stopped next to. The 50-Cal was lighting up this little helpless mud thatch. From my angle, I could see the tracer rounds going in one side and out the other as if the rounds weren't even hitting anything.

Still no movement and no noise, I guessed someone accidentally bumped the MK19 Grenade Launcher, firing two grenades. Then someone reacted to this by getting trigger-happy on the 50-Cal, firing at anything they might be able to see or at least blame. Once this idea made sense to me I felt dumb for my initial reaction, and fell back from my position to my truck.

There was a driver's meeting later that day at 1800 where we found out we would be leaving again at 1900. But at 1930, our serial leader informed us we would be there for three more hours. So Oyster, Scheider and I napped for a few hours and then much to our dismay, we left at 2215 … 15 minutes early. This was the closest to leaving on time we had been since we left Matilda.

Chapter Twelve

An Nasiriyah

Although the original plan had been to stay south of An Nasiriyah for the night, the call had just been made for us to go through the outskirts of the city tonight.

The convoy was slow as usual, traveling about five miles per hour for a few miles and then stopping for as long as two hours for reasons unknown.

Oyster and Scheider were both napping as I watched the lights from the war over the horizon. I could only pray that our guys were surviving. Earlier today we had heard the horrific news that nine soldiers and a severely beaten and raped female were executed on closed-circuit Iraqi television. We then heard it might have been shown on CNN. We also heard that RCT-1 had been overrun and the Iraqis were calling the Americans and our military cowards and pussies.

We also learned of a defensive strategy being used by the Iraqi military. Their combatants would hide and allow our assault units with tanks, AAVs, LAVs, and other big guns to roll through. Then they would fire upon the support convoys as they traveled through the area that had been deemed all clear. The tactic was meant to weaken our forces by cutting off supply and support routes, boosting the morale and heightening the spirits of the Iraqi military.

Finally the convoy began moving again. Oyster was still asleep and the convoy was leaving us since he was in the driver's seat. I woke him up and we played catch-up as Velasquez followed behind us.

A couple of miles later we were stopped again, and Velasquez ran up to our vehicle to let us know there was no one behind him. Apparently Oyster hadn't been the only one sleeping; two-thirds of the convoy was still back at our last stop asleep. I radioed to our serial commander that not everyone was up with the convoy. Eventually everyone caught up and we made our way north.

March 25. The morning glow was now upon us as we had traveled through the night. We would have to enter An Nasiriyah in the daylight.

We received the intelligence that for the three-mile stretch between the two bridges leading into and out of the city, there was still fighting going on. We would pass through in Rifle Condition 1, which meant magazine loaded, round in chamber, ejection port cover closed, and weapon on safe. I was already in this condition and had been for quite some time. And since Marines had reportedly already taken the west side of the road and that side no longer posed a threat, our weapons were to face east.

We rolled into the edge of the city, passing many miles of trash and garbage from an untended landfill. Trash was blowing all over the place.

Then to our left was a piece of destroyed Iraqi weaponry. From the best I could tell, it looked like a BMP-1 Iraqi tank that had been blown clear across a four-lane highway on its side and burnt to a crisp. We could see the helicopters hovering over the city ahead of us and smoke rising up to them.

Across the first bridge into the city were seven Iraqi tanks that had been destroyed in their hiding spots with barrels aiming south at anyone trying to cross the bridge. I was relieved to see this as opposed to our own vehicles, but nonetheless, I was hanging out the window of the truck, alert. I still had the rifle's safety on knowing that if I did see anyone with a weapon I would have to pause briefly to flip the safety, giving me a split second to analyze the threat.

Marines lay in the median in the prone position aiming at either side of the road. So much for the west side being cleared. LAVs, AAVs, tanks, and helicopters aid them.

The fighting continued as we passed. The safety was off in an instant once the shooting began. This was a matter of life and death, and the slightest hesitation or lack of accuracy could have had an adverse effect on my future. Sound became muffled and movements became precise. My heartbeat raced but I felt calm and collected. My breathing became deeper; I sighted and solidly squeezed the trigger.

Many shots were fired at us, but many more came from us. Marksmanship pretty much goes out the window, literally, when returning fire. I can go 10 for 10 at 500 yards on a rifle range with no problem, but I doubt I could go 1 for 10 at 30 yards when playing a form of Whack-A-Mole with an M16 during a drive-by.

Even with my Marine training, I had sometimes wondered about having to shoot someone if this type of situation were ever to occur.

Would I be able to do it? It was crazy when I thought there was a chance I wouldn't walk away from this. It was crazy when I thought I might never again get to see the loved ones I comforted when I was leaving just over a month ago. I realized quickly that pulling the trigger might determine my survival as well as that of the Marines on either side of me.

This wasn't a time to ask questions. This was the time to put my training to use. This was about survival. There are no regrets when it comes to survival, and if someone made a sudden movement that could have a negative impact on my life then if it's up to me it's not going to be my life that is taken. Thankfully, I wasn't hit, and honestly I would like to believe that none of my shots caused anyone harm.

Moments later, we were through the city. We passed four destroyed AAVs—some worse than others. The first was caved in from the back like someone had shot it with a Rocket Propelled Grenade or RPG.

The second was hit by a round from a tank on the top near starboard side, which is where the fuel cell is located. The coolant towers and plenum were blown about forty yards away. Marines were now using them as cover to help shield them in their firefight.

The third was probably hit by an RPG on its tracks and was demobilized. They were now using this vehicle as a parts vehicle for other AAVs.

The last was the worst of all. It was very disturbing to see this AAV and to know that no one could have survived the blow it took. While the other vehicles could have been abandoned, this one could not. The MK19 and the 50-cal are normally taken off before abandonment, but on this vehicle they were still mounted. Black rubber marks led to the vehicle from across the four-lane road and curbed 10 ft. wide median. The top third of the vehicle was gone, and the entire remains were charred black from fire.

Everything happened so quickly out here. I didn't really have time to absorb all that I saw. We crossed the bridge out of the city, as the Cobras were still lighting up random targets. Smoke arose from all over.

A few hundred yards after the bridge, the road ended. We had to turn left or right. The serial commander wasn't sure which way to turn, so he stopped. The other two-thirds of our convoy was still in the city. Even my vehicle was in danger. No one could believe what was happening; we were just sitting ducks. It felt like we had been driving a car down the road not paying attention and then looked up to see another car at a dead stop just ahead. We slammed on the brakes knowing a collision was eminent, yet were powerless to do anything about it. But just before this imaginary collision, the serial commander turned left and we were on the move again before anyone got blown up.

About a mile further, we came to some sort of military compound with a statue of Saddam standing above the gates. Where the statue had once grasped the Iraqi flag, was now flying Old Glory.

In the street in front of it was a Mercedes flatbed work truck with bullet holes in the windshield and blood spattered across the back window. All we could see in passing of the driver was his mangled head lying on the windowsill.

Just outside the truck was another corpse of a man that apparently had been shot in the middle of convoy traffic. The corpse had been flattened repeatedly by heavy equipment. A stray dog feasted on him and carried off an arm as we passed.

Further along was a white and orange Suburban decimated by hundreds of 50-cal rounds. Blood was splattered across what glass was left of the windshield. The blood continued to drain out, building a pool on the street underneath the vehicle. In passing, I could see one guy twisted and outstretched from the front passenger seat to the rear driver's side seat with his broken leg draped out the rear door.

Just a bit further up the road was another dog. He was dragging the lower part of a human leg across the street. We all looked around for the rest of the body, but it wasn't in sight. There must have been more bodies in the compound, or maybe this was just the last piece of someone.

Continuing further north, Iraqis with white flags waved us on. The somber looks on their faces left us to wonder about their reasoning for waving the white flags. Were they supporting us or fearing us? It was hard to understand what anyone was feeling now, if anything at all.

Then I began to question my own mental well-being. Here I was, taking pictures of death and carnage that modern day war movies don't even do justice. I witnessed Marines turning already dead bodies into road kill and stray dogs eating fresh human remains. There was literally no respect for the dead.

I wasn't acting the way I thought I would. But how did I expect to act? I mean, I could tell myself all day long that this was the type of thing that I would see but until it happened, there was really no way of knowing how I would actually react. I guess I thought my stomach would be turned or that I'd want to look away. Quite the opposite.

I tried to pawn it off as nothing, and immediately began to think of all the people I would not tell as well as the people that I would tell if asked and how much I would tell. Abby had made me promise to tell her everything, but I decided I could break that promise. There was no way I was going to tell Mom about any of this. Dad, David, and Kim are on the "Tell if asked" list but I didn't have a "Tell" list because I understood many would find these stories disturbing.

Suddenly, there came a strong sandstorm. We closed our windows to keep from getting sand in our eyes. At the same time, we had to mask up because of a possible NBC attack. We were stopped in front of an unsecured military outpost, so we set out security. This was ineffective though; the Marines were just lying there on the ground with their eyes closed because the sand was blowing so fiercely. Then it began to thunder and rain, so the troops came back in. The Tracks kept their positions guarding the flanks. The storm quickly passed, and nothing came of the NBC risk.

And just like that, everything was back to normal. A dog pranced by my vehicle and I tried to call him over. I guess he didn't understand English and I wasn't speaking Arabic, so he continued on. Or maybe he was in search of more food?

Finally, we were on the move again. But we hadn't gone more than a quarter of a mile before we stopped once more. We were stopping so many times during these convoys it was annoying. But then the same dog came prancing by and this time I had food. I tried to persuade him with some Mexican rice out of my MRE but even he wasn't having any of that stuff.

In my attempt to lure the dog near with the door open, a magazine clip fell out of the truck. As I jumped down to retrieve it, a gust of wind swept through and blew the inflatable pillow Abby had sent to me out of the truck and carried it away. It felt like it was a piece of Abby slipping away, and it terrified me.

This gust of wind was the first of many in a new sandstorm beginning to come through. I sprinted after the pillow with my rifle in the alert position. The wind must have been blowing 40 MPH. I knew the pillow was gone. It was flying in the direction of another unsecured area. My life was at risk if there was any resistance ahead.

Yet here I went, sprinting across the street with my Kevlar helmet bouncing up and down on my skull and the pillow clearly blowing in the wind faster than I could ever run. Once it stopped to let me get close, only to be swept away again as I reached out for it. It finally rested just long enough for me to catch up to it, and I dove on top of it as if it were a football lost in a fumble.

Everyone who witnessed it got quite a kick out of this sight, and I would never hear the end of it. Granted, the way I looked chasing an inflatable pillow across enemy territory must have been pretty funny. But even so, I couldn't have been more relieved to have saved it. Scheider and Oyster just stared at me as I climbed back inside the truck.

Unfortunately, the fun would be short lived. As we started moving again we came upon a civilian city bus. It had obviously been blown up but there were two Iraqi men standing by the road handcuffed to one another. One was wearing only boxers, and I could tell he had been in the explosion. His ears were bleeding and he had no equilibrium and no balance to stand all the way up. He was burned so badly that his skin dangled from his body and he was shaking uncontrollably. His face expressed his agonizing pain and he didn't have a clue as to what was going on.

The man cuffed to him, also burned and wounded, tried to get away from the burning bus but all he could do was watch his helpless friend continue to fall and drag him down. I wondered what was going on in their minds. Surely they were offered help before they were handcuffed together. Maybe they refused medical treatment and had to be stripped of their weapons and cuffed in order to keep them from being a threat.

On the opposite side of the road, beside the burning bus, was a pile of a dozen or so bodies. The body on the top of the pile was only half the size of the average man. Maybe I didn't get a good look since we were passing so quickly. I wasn't going to let myself believe what I had just seen. Maybe it was a torso, or maybe it was just a short man?

Another distraction took my mind off the thought. Up ahead on the left was a man missing the lower half of his body from the waist down. He was dragging himself across a sandy clearing away from the burning vehicles. His insides and intestines were stretched out behind him from where he started, but his legs weren't in sight. Was he alive during this escape, or was it pure adrenaline carrying a dead man? How far would he make it before his final resting point?

Near the road was a decapitated body with its head some distance away. Up to this point I had been keeping an accurate count of what I was seeing. But now the numbers began to seem relentless. I had seen headless bodies, limbless torsos, and gutless insides all while watching a select few survivors suffer to death.

What got to me the most was the stench of burning flesh. They say that smell is the most memory provoking of the senses. I hoped I never had to smell burning skin or think about this place ever again.

A few miles after this, at about 1500, we met up with other assault units who had just blown up a military compound approximately 50 yards from the road. We would be staying there overnight. The building was still burning and would continue to burn throughout the night, along with the bodies inside.

The smell permeated the air and made ordinary tasks hard to do. I struggled to get through them. The simplest of things was difficult. Like talking, for instance. When I opened my mouth, I could taste the smell. Religious and spiritually haunting thoughts entered my mind as the contaminated air I was breathing entered my body.

I tried to eat that night, but didn't make it through the first bite of a pound cake pastry. If I was going to be able to eat, pound cake would have been my best bet. Not only was it the best tasting pastry, it was solid and dry. I couldn't imagine trying to eat something mushy, or in a sauce.

My roommate in the barracks back at Camp Lejeune, Cpl Moya, came over to me to talk for a little while. He'd had the chance to take a closer look at what I had seen but refused to believe earlier.

"Hey Cox, did you see that bus we passed earlier?" Moya questioned.

"It was crazy, man," I replied.

"Did you see that kid on the top of the pile of bodies?" he asked.

"Well, I hoped I didn't see that. Is that really what it was?" I asked.

"Yeah Dude, it was a little boy. Couldn't have been older than 8," he confirmed.

"How can someone do that to their own people?"

"Seriously. What a fucking asshole!"

This must have been the most eventful day of my life, for all the wrong reasons. It was time for me to go to sleep, then, so I zipped my sleeping bag all the way up hoping the smell wouldn't be so bad. Plus I didn't want the haunting spirits resting in me.

I was woken by an explosion just about eighty yards away from my truck. Then another explosion, a lot closer. We were being fired on by long-range artillery and there was nothing I could do except pray that our counter-artillery battery would take them out before they took us out. The next round landed just a bit further away than the last, but still closer than the first. I zipped my bag up again and closed my eyes.

If I'm going to die tonight, then I'd better hurry up and go to sleep. People question the way they want to die. I want to die at peace in my sleep. I'm a heavy sleeper, so I won't wake up for my death.

Chapter Thirteen

The next morning, March 26, I woke and discovered we had taken many EPWs overnight. With so many, we began using our AAVs as Enemy Prisoner of War transporters. We could fit about twenty-five in the rear troop compartment. It took three vehicles.

Some suspicious-looking Iraqis were in the area, so we were on the move again by 0800.

Shortly thereafter, we stopped again. This time we were beside another bus that had been blown up from behind. It left seven dead but had spared the driver, who was leaning against the front driver's side tire. His right side was burnt from the explosion and he had a deep wound near his kidney bleeding profusely. His right eye was slightly melted and oozing.

He looked up at us with his one left eye. He pointed to my rifle that was aimed out the window at him, and made a hand gesture like he was pulling a trigger. He had just enough strength left to get up and walk toward us while repeating the same hand gesture—pointing at my rifle and squeezing the imaginary trigger in hopes of putting an end to his suffering. I radioed in to ask permission to fire but the order was given to fire only if threatened. He grabbed at the end of my rifle in desperation to get what he wanted. I got out of the truck and forced him to sit back down as if I was speaking to an animal.

I decided to walk around the bus and examine the bodies. It was hard to make out what was what. All that was left was mostly raw flesh. Then I made a shocking discovery. Amongst the flesh was the burnt skin of a woman's breast slightly attached to her body. It was painful to see this sort of thing, along with the child yesterday.

There was a man lying on the other side of the bus that had been blown out the side during the explosion. His legs were broken, intestines spilled, skull crushed, and his brain was oozing out. His face held the look of pain and terror—the last thing he felt before his life was taken from him. Beside him lay hundreds of 7.62 mm rounds for his AK47 that would have been used to kill us.

If the Iraqi military would just understand that we will not harm them if they do not resist us, this would all be so simple. We would treat them better than they are treated here, and provide medical treatment and food. But they insist on fighting a force they cannot compare to.

We climbed back into the truck and were about to move out. Looking down at the bus driver, his eyes were now closed. His shaking had almost stopped and his breathing slowed. He opened his eye one last time and gazed upon the earth. Shutting it, his shaking stopped as he released his last breath.

Further up the road, more vehicles were blown up and more bodies were scattered. The vehicles were still smoking and the stench of burning flesh enveloped us once more. Another bus was still smoldering. The driver was blown out his door. In the midst of the commotion, the windshield wipers of the bus had been turned on. The screeching from the wipers scraping the glass made a haunting sound. The repetitions faded slower and slower as the battery was slowly dying. The destroyed vehicle outlived its passengers, but the slowing sound of the wipers was like a heartbeat fading away.

I took a short nap, as it was nearly dark. When I awoke, it was so dark I couldn't see my hand in front of my face. The NVGs were working for only about five feet, which was just enough to see the end of the truck's hood. NVGs work by amplifying pre-existing light so they only work when some light is available. On top of this, we were lost from the vehicle in front of us, and Scheider was flipping out. This made for an annoying ride, but then Oyster volunteered to drive for a little while. I did what I could to guide him with the NVGs, and we eventually caught up and were staged in position for the night.

March 27. We woke to the sounds of helicopter and tank fire just ahead. We learned that in the blackness of night, we overshot our staging point and stopped about five hundred meters from a small hostile town. The Iraqis had the road guarded with tanks and artillery. More than a time or two during our traveling, someone had accidentally turned on a light at night. Had this happened last night, our position would have been given to their defensive assault units and we would have encountered hell. Fortunately for us, at daybreak our guys spotted their positions and spent a couple of hours blowing them up.

During this time, I helped watch the EPWs. We had to piss, feed, and doctor them. It's unusual and embarrassing to have to do this sort of thing with grown men.

After all this, we continued to sit on the side of the road. I happened to be outside of my vehicle when I heard Ryan calling my name. He and Russ were at the truck behind ours getting oil.

We talked for about fifteen minutes just recapping what we had seen and gone through in the past couple of days. We traveled in the same convoy, but didn't get to see each other that often. They were in the back of the convoy while I was further up past the middle, so there was about a mile separating us. I usually only saw them when a BSA was set up.

A little over an hour after I had spoken with them, we mounted up and moved out toward our BSA. Two hundred meters from where we started, we stopped again. Apparently this was our position and we had wasted an entire day parked on the side of the road doing nothing when we could have been repairing vehicles.

Ryan and Russ were even more upset than I was. They had just walked a mile to retrieve two five-gallon jugs of oil each for some Tracks that may have been slightly low. Not knowing how far to the next BSA, they thought it best to fill up on oil before marching a long distance. They returned to the Tracks exhausted from carrying the eighty pounds of oil each. To make matters worse for them mentally, we advanced forward to where they could park beside the vehicle they had just journeyed to.

Our moods changed that night as they normally did when things quietened down and we had some time to be alone. I'd been sleeping up on top of the camy netting in between two metal quadcons on the bed of the truck. These quadcons measure about 8' x 5' x 6', and have barn doors on either side. As large as they are, they make good field offices and one of ours served as our tool room. Where I slept was about 12 ft. off the ground; the camy netting was on top of two resin convex boxes being used to store spare parts.

I reach my sleeping position by climbing a ladder up to the tool room quadcon. I then stand inside the quadcon, face outward, and grip the edge of the quadcon's roof. Hanging, I then curl myself and flip my body up, back, and over the roof. Now on my stomach on top of the quadcon, I push up with my arms and I'm there.

From my position, I could see stars for the first time in a few nights. When I can see the stars, I always look for satellites or shooting stars. I had seen neither on this night, but I couldn't help but notice one star so bright it seemed to make the others look faint. It made me think of Abby and how the two of them were similar. Abby has a way about her that makes everyone else in the world seem a little less visible in my eyes, just like the other stars seemed faint in the light of this Star.

I began singing that "Wishing on a Star" song, but I couldn't remember many words to it so I made up my own lyrics. I asked the Star to tell Abby and Mom that I was O.K. and not to worry about me so much. I wished I was in Abby's bed with her wrapped in my arms. I wished I could see her smile and her beautiful eyes. I missed the way she bites her bottom lip while looking into my eyes the way I was doing then with my eyes tearing up. Just as I was making my wishes, a star shot across the sky. I drifted off to sleep with a smile.

During the night I got cold. My MOPP suit, sleeping bag, and poncho liner weren't enough to keep me warm. I woke up shivering at 0100, 0230, and 0330. Finally I got up and climbed into the cab of the truck try to warm up and eat. Ryan stopped by for a little while as I ate but my breakfast was cut short.

March 28. There was work to be done on two vehicles. We had to take off seven good road wheels, three support assemblies, six shock absorbers, and swap out cooling towers with a third vehicle. I worked with Russ for a little while, but then he had to leave to go pull a "pack"—an engine and transmission assembly.

I quickly found myself working on Ryan's vehicle, directing him as he operated the boom. He and I were able to talk during breaks. We took bets on how long we would be here in Iraq and when we thought we'd be back home. He seemed to think we'd be back in Kuwait or at least out of Iraq by April 10. My guess was the 18th, but this was one bet I really hoped to lose. With the way things were going, I couldn't help but think we would be here for at least another three weeks.

As we were working, they had to segregate the EPW officers from enlisted and search them for weapons. There were about seventy-five men, and apparently there was a knife found on one of them.

There are many mixed feelings about the EPWs out here. The Marines that had spent time looking after them treated the EPWs with respect, offering them cigarettes and whatnot. But other Marines, not having spent any time with them, would see these acts of humanity and yell that the EPWs were the same men who had been trying to kill us.

As for me, I remained neutral. I can see how someone could develop compassion for another human being after having to watch over them as if they were a pet. Having to feed and water them, piss and shit them, undo and redo their pants, take them to a medic to remove bullets from their bodies, and so on. Yet I can also understand those who feel a great hatred toward anyone that would point a weapon in our direction with the intent to kill.

At about 1400, I went back to my truck where we now had a fire burning our trash. I pulled up a box of MREs to use as a stool. I ate my supper and smoked the last day's cigarette by the warmth of the fire.

Unfortunately, I then came to find out that I had security watch on the 50-cal ring mount on top of the LVS. So I retired early in hopes of getting some sleep before my post, but not before counting four satellites and wishing on my Star.

At 2330, I woke up freezing. I tried to warm myself but it's difficult at night. It had gotten windy and blown one of my socks off to the ground. And where I slept, it wasn't like I could just hop down and get it. I had to go through the same steps it took me to get up, only in reverse. So I was glad when LCpl Colon came by a few minutes later to wake me for watch and he tossed my sock up to me.

I took my blanket with me to my post, but it did little good. In the wind and bitter cold, my toes became numb. When I was relieved, I felt frozen to the bone and had trouble moving. It might seem that it would begin to warm when the morning glow appeared in the sky, but for some reason, this was when it became the coldest and the wind became even stronger.

I decided to build another trash fire and shortly afterwards, was warming my numb toes by its side. Just a little later, word came down that outgoing mail would be sent off. I quickly began writing a letter to Abby. I had so much I wanted to tell her but for some reason as I tried to write, I couldn't find the right words and said the same things over and over again. I tried to sound busy, but all I could think of was the death and carnage we had just seen.

About that time, SSgt Little came by and asked for three volunteers to go on a patrol through a hostile village where contact was imminent. I quickly volunteered, along with the two Marines sitting next to me. This seemed like a chance to experience a once-in-a-lifetime event to gain intelligence and really put my training to the ultimate test. After having these thoughts, I questioned my own sanity. I hurriedly finished my letter to Abby and got ready. I couldn't help but wonder as I was mailing the letter if it would be the last one I sent, but then laughed off the thought.

March 28

Abby,

Hey Baby! I'm alive! Ha ha ... I'm somewhere on the side of the road about 120 miles out of Baghdad. They ceased all incoming and outgoing mail, so this is the first I've been able to get word out since we left Matilda on the 18th. All our incoming mail is just piling up somewhere in Kuwait. I can't wait until I can get it.

You wouldn't believe the things I have seen out here. I promised to tell you everything I've experienced, but after I've encountered these things, it may be difficult for me to tell you or anyone. But a promise is a promise, so I will tell you only if you really want to know. I've been keeping a journal that doesn't leave anything out so if you really want to experience it, you can read that. The thought even crossed my mind to write a book when I get home, but that's pretty far out being that I don't like to read.

Sorry Babe, but I have to cut it short. I have to go on a patrol to clear a small village. Wish me luck. Tell Mom and Dad that I am O.K. I'm not going to have time to write them. I hope that things are going good for you in Charlotte. Tell everyone I say Hey. I can't wait to come home. But our toughest part is still ahead— Baghdad. I love you so much! I have a star in the sky picked out for you. I talk to it every night, and I tell it to tell you that I am O.K. I hope it works. I'll write again whenever I have another chance. Please help calm down my Mom. She has to be a nervous wreck, especially if the rumors are true about what they have been saying on TV. Tell them I'm sorry for not being able to write them too.

I love you always,
Eric

At 0700, we were in front of the Command Operational Center, or COC, being briefed on the mission objective. I quickly realized my hopes for a real adventure had been misguided. The task of going house-to-house clearing a small village quickly changed to going across the wheat field beside us to two houses and asking them if they knew anything about the military. Even if they had known anything, we wouldn't find it out because they were speaking Arabic and we were only there to provide security, if needed.

And that is exactly what we did. At 0900, we returned from our pointless waste of time. For the remainder of the day, we sat around doing nothing but talking about different rumors everyone had heard.

PFC Prince heard from MGY Reece that we would be here at this BSA for another six days and there was a chance for incoming mail and hot chow. This was news to all of us, but we hoped the six day part wasn't true. The last we had heard from our Company Commander, Captain Ferris, was that we would be leaving at 0330 the next morning.

Another rumor was that the American female we had heard had been raped and executed was actually a Marine and not a Soldier, and that she was still alive. Also, not all of the American POWs were executed, but the ones that were had been shot in the stomach to die slowly and painfully as the others watched and waited their destiny.

Then we heard that a Marine Sergeant was left behind in a town about thirty miles south of where we had just come from. Supposedly he was killed, hung, and put on display in the middle of the town.

News of this nature spread quickly. Most Marines refused to believe these things and cast them off as nothing more than rumors. Some even went so far as to say these were strategic rumors designed to brainwash us into becoming cold-blooded killers. Sadly, many believed it to be working.

We began telling jokes and laughing to change the mood. Shortly after came nightfall, and most retired to sleep.

I, on the other hand, got stuck with security watch again on the 50-cal ring mount. When I looked at the roster to see who I would be waking up to relieve me, I realized that the Corporal of the Guard or COG had screwed up the watch roster.

I spent the first hour of my post looking for Epperly of Motor T but he was nowhere to be found. He was in charge of making and regulating the watch roster. I then learned from Cpl Duncan that Epperly was in a fighting hole until 2100.

The fighting holes were spread out around the perimeter of our BSA. At night, someone had to point us in the specific direction of their location, or we would be lost. Even worse, we ran the chance of going outside our perimeter and might be mistaken as an enemy intruder.

I wasn't about to go looking for him, so I woke up LCpl Ballos, who was next on the roster at 2400. I told him that he needed to find Epperly at 2100 when he was relieved to fix his mistake on the roster.

Shortly after my relief at 2030, I was fast asleep. I slept in the next morning until around 0400.

March 29. At 0430, we gathered for word from CWO3 Perser. He tried to play the role of a father figure by telling us what a great job we were doing and how thankful he was that we had taken no casualties. But he has a habit of uttering "mmmkay" after almost every sentence. It's so bad that none of us really heard much of what he was saying because we were all too busy counting the number of times he repeated it. And then there's his arm scratch. He crosses his arms when he talks and instead of tucking both hands, one of his hands is constantly scratching the other forearm. Throughout his entire speech, it was like he was holding and petting a pretend cat!

Unfortunately, we did learn that the awful rumor we heard last night about the Marine being put on display was in fact true. This greatly disturbed all of us. We were relieved to hear that tanks, LAVs and air support were going back to retrieve the body, and that their orders were to then level the town and destroy anything that moved.

It was also stressed not to become complacent and let our guard down after we had come so far and been through so much. NBC awareness should be heightened as we neared Baghdad.

The town of Al Kut would be avoided in hopes of pulling their security north and behind us. This way, we could engage them away from their concealed positions, thus minimizing casualties and fatalities.

It was brought to our attention that the death and carnage we had seen had been masked with civilian attire and equipment. It was assumed that most of their military assets had been pulled in to defend Baghdad.

Finishing the meeting on a positive note, we learned we had over 3,100 EPWs and had destroyed numerous companies, batteries, and battalions of their army.

March 30. While making my "Wake me up" drink, I was instructed to report to the COC along with Moya and Norrie—the same two Marines I went patrolling with yesterday—in fifteen minutes. I knew exactly what this was for and quickly asked if it had to be me or if someone who hadn't been on a patrol could go in my place. But no … they wanted the same eighteen that went the day before.

I took my time getting ready, in hopes of getting out of it somehow. Thirty minutes later, they sent Moya over to get me. Fifteen minutes later at 0800, we were in a tactical column traveling the same route we had traveled yesterday, only in reverse.

The first Iraqis we came upon were five shepherds in the wheat field. Talibe, our Free Iraqi Forces translator, spoke to them along with Major Bailey and a Lieutenant Colonel.

At the end of the conversation, we gave them a box of Humanitarian Rations. But they didn't trust us or our gift. Talibe and Major Bailey each broke off a piece of toaster pastry, and each ate the piece. Then they handed it over to the spokesman of the five. This "spokesman" had been the one more willing to talk, and the other four seemed to do as he did. He broke off a piece and handed the pastry to me. I did the same, and handed it over to the Iraqi in front of me. We all ate our piece and were answered with "krun," which supposedly means "thanks" in Arabic.

After we said our goodbyes we headed in the direction of the house we had visited the day before. While walking, I couldn't help think about that word "krun." I mean, for all I knew, it could have meant any number of things and I could have responded with any number of replies. The possibilities humored me as we neared the thatch.

Two dogs came to meet us barking ferociously. But then they seemed to sense that picking a fight with eighteen combat-ready Marines probably wasn't in their best interests, and retreated to the wheat field behind the thatch.

When I say "thatch," I mean this is a four-bedroom, no bath clay thatch of a house. It did have special features, like bonus rooms for goats and chickens as well as open-air hallways. There were also detached family rooms of some sort that had straw laid on the earth in place of carpet.

It was sad to see. A family of about seven lived here—a man, a covered woman, and about five children that ranged in age from newborn to no older than ten. We had to clear their home and see if anything had changed since yesterday.

A cute little white puppy was in one of the rooms and ran to hide in another smaller room. I of course followed and tried to call it out of its' corner. He couldn't have been more than a few weeks old. He was wagging his tail like he wanted to prove he was friendly and would have liked so much to come out and play. But he was trembling and shaking, and then I realized he might have been emotionally scarred by something that happened in the past week.

I wanted so badly to rescue this little guy, but I knew it wouldn't fly with my Company. I also knew that if I showed him an ounce of compassion he wouldn't leave my side. I thought it best for me to leave him alone and go on about our mission.

We moved out to the second house that had a stagnant mote with all kinds of insects looming. They talked to the same people from yesterday, and then we patrolled back to our position.

Upon our return, we were tasked with opening up an engine container to see if parts were good. Busy work, basically, but it helped get my mind off the little puppy and the people we had talked to. Then we cleaned our area, as the trash had begun to accumulate.

Later that night, I played a movie game with Nik, although I had been angry with him earlier when he had successfully skated the clean-up festivities and I had been left with cleaning his entire area. In a skateboarding magazine I picked up during the clean-up, I found some cool shirts and beanies I wanted to buy when I got home.

Night fell, and by 1630 I was in my sleeping bag. It was the first night I slept without waking up cold. I was, however, woken up by the artillery being fired around us. The sound was somehow comforting. I thought to myself, "Someone's getting lit up!" And then I drifted back to sleep.

March 31. At 0330, I was up and making one of my various drink mixes. I traded out some items I didn't much care for to get the good mixes like cocoa powder, coffee, spiced cider, and beverage-base powder. The cider drink is just cider, the beverage base is just a flavor like grape or lime, and then I mix the cocoa, coffee, sugars and creamers to make my "Wake me up" drink which is good at any temperature.

I decided to take a nap around 1200, but not thirty minutes into it, I was woken up to help fix broken Tracks. Whenever we got broken vehicles in, there were usually items left behind on them. I slowly made my way over and began looking for articles that had been discarded. I had lost my beanie several days earlier and was scavenging for a new one. I only found a Mag-Lite, but it was a keeper. I ended up chatting with Ryan and Russ more than I worked, but I did work.

At 1530 I ate my last meal of the day—a meatloaf and mashed potatoes with gravy MRE. This is one of my favorite menus, but it sure didn't come close to the way Mom makes it.

It's time for me to head off to my sleeping bag hiding spot. I love my sleeping spot, because no one thinks to look for me up on top of some storage containers on the back of a truck.

Chapter Fourteen

April 1. I woke up chilly at 0230 because I was using my extra blanket as a pillow. We were scheduled to advance fourteen clicks north to northwest at 0500. We had been sitting at BSA-3 for five days waiting to move, and on Day Six our wish was finally granted. At 0510, we moved out.

Shitter had just said, "It's scary that we're actually moving!"

"I'm kind of excited!" I had answered with enthusiasm. But we didn't move quite as far as we had expected.

Our excitement quickly turned to dismay as we watched the quartering party pull off the road across from where they had just left and begin setting up BSA-4. It was not even 0515 yet! We began looking and scanning the area for some sort of logical explanation.

It was now 0645, and the only logical excuse for such a movement was that today is April Fool's Day, and the joke was on us. On the other hand, every day here seemed like April Fool's Day. I thought about the one positive aspect of the day—since it was the first of the month, I got another $300 of combat pay.

To boost my low spirits, I thought back to my younger days at our cottage on Badin Lake. It was on the main water channel but the way the lot next door juts out into the water causes debris to wash up at the corner of our lot. I would wake up early in the morning to go down to our pier and look for "treasures," as I called them, which may have collected during the night. They were generally fishing lures or small toys.

But these days were a little different. As we pulled into BSA-4, I had noticed some sort of fabric or clothing lying on the shoulder of the road. After we had parked, I went on a journey to retrieve whatever it was I had spotted.

On the walk over to the general location of this "treasure," I found unused packets of coffee, sugar, creamer, spiced cider, and other usable items. Free mixes I didn't have to trade for!

Then I came to the fabric I had spotted. I stooped over and realized it was an Iraqi uniform. I picked up the Iraqi blouse, and guessed whoever had thrown it away had done so in hopes of not being shot.

I walked further and found a Camelbak drinking system with a hole in the bladder that some Marine had thrown out. I picked up more packets of coffee here, too.

Continuing on my path of fortune, I came to a complete khaki Iraqi uniform and a set of keys. I had walked as far as I wanted and my hands were full, so I started back.

Then I spotted something lying in the bushes. I couldn't believe it when I found it was a complete green Iraqi uniform with a belt. A red rope was tied around the shoulder, probably denoting an affiliation with a particular unit or perhaps an award.

Returning to the truck was like "show and tell." Everyone was constantly looking for souvenirs out here. I packed my treasures into a plastic trash bag and hid them in a quadcon near my pack.

Knowing there was still unexplored territory, I decided I was not finished with my quest. I departed again, exploring the small area I had missed on the previous expedition. This time I found less, but more than accomplished my mission.

I stumbled upon a poster-board sign that could be of interest. I turned it over and realized it was of *major* interest. I had found a poster of Saddam Hussein. He was smiling and his hands were clasped together in a praying manner. Behind him was the red, white, black and green Iraqi flag and the phrase, "Allah Akbar'" which I guessed had something to do with God. There was another scripture at the bottom written in Arabic.

Now my journey was to find Talibe to have him translate the meaning. He explained to me that "Allah Akbar" is Arabic for "God is Great." The Arabic writing at the bottom translated to, "The Honorable Mr. Saddam Hussein 'God Bless Him.'"

I went on tour with my new finding and ran into Master Guns Ogilvy. He was in the middle of digging a hole, so I just stood there holding the poster facing him as I waited for him to look in my direction. Shoveling another scoop of dirt, he looked up and saw the poster. Without skipping a beat, he threw the shovel full of dirt against the poster of Saddam.

"Are you one of his followers?" he asked.

"You could say that," I replied. "Following him with orders to kill him!"

After a few laughs, I left to show Russ but found that Russ had an even more important story. Apparently, Sgt Chavez had lost control and was plotting to kill Russ and SSgt Little:

When we had pulled into this BSA, Little and Russ were discussing our travel situation. Little was pissed and was cussing up a storm. "We're fucking flying!" he muttered sarcastically.

Chavez, hearing Little's remark, walked up to him and said in his small, squeaky voice, "Well, you know, Staff Sergeant, we're actually moving pretty fast."

Little looked back at Chavez in disbelief and replied, "Yeah Chavez, we're fucking flying!"

Chavez, realizing what he had first said came out wrong, tried to correct himself by saying, "Staff Sergeant, what I was trying to say was ..."

"No, you're right, Chavez," Little butted in. "We're going to be there any fucking minute now!" as he looked at his watch.

Chavez got quiet and cowered down. Little had just made him look like a fool and now Russ was laughing. Chavez, without a word, turned and walked away.

Moments later, Chavez returned with his E-tool and began digging. He finished digging a rectangular hole, about 6 ft. by 3 ft. and 4 ft. deep. He then started digging a second hole, and Russ and Little began to wonder what he was doing. They sat there and watched, trying to figure it out as they continued to talk about him and laugh. But then Little suddenly stopped talking in mid-sentence and turned to Russ ...

"He's digging our graves!" Little announced as if he had just solved a riddle.

"That ain't cool ..." Russ answered. "We've got live rounds out here!"

Chavez was immediately sent for counseling.

At around 1200, SSgt Lopez needed help moving parts out of the towed vehicles because we were giving them up later on that day. After this transfer was complete, all of the "2141" mechanics got an ass chewing for not working like we should have. The lecture was not unusual; this type of thing happened on a regular basis and was taken with a grain of salt.

After the ass chewing was over, I went back to my truck to rework my drink recipe. There was way too much coffee and not enough sugar or cocoa. Since I was out of cocoa, I just added sugar and creamer. It was getting better, but still needed improvement.

The day could not have been passing any more slowly. I found some clippers and decided to shave my head again. I bent over as far as I could in hopes of not getting little pieces of hair all over me; it had been fifteen days since I'd had a shower and the last thing I needed was to be even more itchy and scratchy. Nik made sure I didn't miss any spots, and I returned the favor for him.

It felt good having a fresh new haircut, but somehow it made me miss Abby even more because she was not there to see it. She always noticed any little change in my appearance. I went to bed that night in the normal fashion except that my wishes to my Star might have been a little more heartfelt. And the kiss goodnight to my ring may have been just a little more meaningful. It was sad to think how such a small thing could make me feel so alone.

The next morning, April 2, I was up at 0300 for accountability before we were to leave at 0400. At 0330 I verbally counseled LCpl McNair for not showing up for morning accountability. Normally it would not have really mattered to me, as almost everyone was going through different stages of rebellion. But he didn't show up for yesterday evening's accountability either, and I didn't want this trend to get out of control. Plus, when we didn't have full accountability, everyone suffered by having to wait around longer.

Moments later, McNair was needed for a welding job and was nowhere to be found. This called for a Platoon-level ass chewing. Apparently we weren't passing word properly and our accountability was poor. Just another grain of salt.

I went back to my truck, took out my book and continued to read. I was determined to finish this book by the time we left here, since for me, it would be a major accomplishment.

I've always hated to read. According to my Mom, even as a toddler I refused to tolerate her efforts to try to read to me. I hated reading so badly that I would even read only the first, middle, and last part of the Cliff's Notes summaries of the books that were required reading in school.

My favorite book was one called *Scary Stories* because it had so many short stories in it. I used this book to write book reports for school. It actually got me through elementary and middle school and even a portion of high school. I would begin the book report saying I had read the entire book but it was difficult to do a report with so many unrelated stories. So to give a better report, I said, I was only going to review the one story in the book that happened to be my favorite. And any one of them was my favorite.

I recently read *The Hobbit*, while standing guard back in the States as the Company Duty NCO. I was proud of my accomplishment, because it was the first book I had read from cover to cover without being forced to do so. And I actually enjoyed it. So to read a second book that has over a thousand pages in it would be big for me.

Just as I finished the "Treebeard" chapter, three CH-53 helicopters came by carrying crates of supplies in nets hanging below them. They hovered over us, sat the crates on the ground, released the ropes, and were gone. It was more ammo, but not for us.

At 1330, we finally rolled out of BSA-4 enroute north toward Baghdad and our next destination. We traveled through small villages where the villagers were coming out to the street to wave and give us thumbs up. We all liked seeing this, as it helped reassure they were looking to us for help. Nine miles from our last BSA, or fifteen clicks on the map, we set up BSA-5.

This spot must have been the worst piece of real estate they could have possibly chosen. As dry as it is out here, we pulled off in what must have been the only mud in the country. No vegetation, just a flat, dead plot of muddy dirt.

And to top it all off, somehow we had wound up right in the middle of someone's farm. Cow pasture was all around us. There was a small neighborhood beyond the pasture. Every thatch must have had at least a dozen barking dogs, and there were about a dozen thatches. And apparently the dogs weren't the only animals.

Even as tired as we were, it was a major challenge trying to get some sleep that night in the midst of what seemed to be over a hundred barking dogs along with the fighting we could hear about six miles north. I laid in my sleeping bag for about an hour before finally drifting off to an uneasy sleep. Then, at 2230, after about three hours, I was woken by the sound of a rooster crowing. I'm not sure if it was all the fighting and explosions going off or if it was all the commotion of our convoy setting up our BSA, but something had screwed up the clock of this particular nearby rooster. I looked at my watch and thought I must have been dreaming, and went back to sleep.

"Cock-a-Doodle-Doo!" It was then 2330, and I was not dreaming. His crowing continued for a while. It was beginning to piss me off. I fell back asleep, but not for long.

"Cock-a-Doodle-Doo!" again at 0115. I was on the verge of taking a nightly stroll to put a bullet down his worthless cock-a-doodle throat. But then he quit and I dozed off again.

"Cock-a-Doodle-Doo!" At 0230, the little shit finally got it right and crowed for the rest of the morning as the sun came up. "Cock-a-Doodle-Doo ..."

Then it got really fun and Old McDonald would've had a blast. After the rooster woke everyone up, the dogs chimed in with their barks. Then came the cows and their mooing. Next the sheep started in with their bleats. The donkeys added their hee-haws ... a sound that must be one of the worst animal sounds in the world. The gnats started swarming up my nose and buzzing in my ears like the annoying little pests they are. So there I sat, right smack dab in the middle of Old Farmer Habibe's Farm!

April 3. We were now on a one-hour notice for departure. But instead of heading north, supposedly we were going back south, hanging a right at our last BSA site and heading west. They didn't think the city north of us had been secured, and don't want us driving through the middle of it.

Going back to try to catch a quick nap after my restless night, I found my pillow, flat. I blew air into it, but I could hear the air escaping almost as fast as I could blow. I found the leak was coming from a 1" slit. I looked for glue or adhesive to patch it, but because of the material it was made of, there was nothing that could be done.

Facing the fact that I no longer had a pillow, I went back to the truck. It really sucked it was gone. I had become attached to it. But on the bright side, it was nice while it lasted.

I sat in the truck, uncomfortable now, waiting for our scheduled departure at 1200. At 1245, it was so hot and the gnats had gotten so bad that I had to get out and walk around.

When I climbed out, I looked down and noticed an anthill. I stomped on it, leaving hundreds of ants trapped inside. A half-dozen or so crawled around the area, seemingly confused about what happened to their home and family.

Having entirely too much time on my hands, I began to feel sorry for the little things. I sat down in the midst of them and helped them dig out their family. Then I helped them rebuild a new farm complete with escape tunnels so someone like me couldn't come by and wipe them out again.

The odometer read 280 miles as we rolled out in the dark, four hours behind schedule. Only this time we were burning our headlights. This was apparently a tactic to lead the combatants in Al Kut to believe we were retreating, which would hopefully pull them out of the city so we could attack their flanks and call in an air strike.

Shots rang out just thirty minutes into our convoy. Two flashes from explosions went off up ahead of us. Tracers were whizzing past us in both directions like laser beams. The scary thing about tracer rounds, other than seeing bullets fired in our direction, is that tracer rounds are placed every fifth round. Between the beams of light were four rounds we could not see.

It seemed like we were caught in the middle of an ambush, but then the firing ceased. I aimed my M16 out of the window but could see nothing. I grabbed the NVGs and scanned the darkness, but still couldn't see anything on either flank. They had to be out there somewhere, but we had to wait for positive identification. I strained my eyes trying to spot something, but there was just nothing.

We came to the intersection of another road and made a right turn. The shots would have been coming from a position just after the right turn in the direction we were going. Still I searched, still finding nothing.

We traveled west for quite some time before pulling over at 2115 to rest. Then we learned there had been a negligent discharge from a vehicle in a convoy ahead of us. Our convoy returned fire as if we were being ambushed, as did they. We were told there had been no casualties in this exchange of friendly fire.

April 4. At 0100 we were moving again, but I was exhausted. I quickly fell asleep, marking the first time I had slept while traveling. I hadn't wanted to miss something by sleeping while we were moving.

I only slept for about an hour before the sun came up. We were northbound on a four-lane highway running about forty miles per hour but having to stop occasionally.

We arrived at the refuel point at 0600, which was our first scheduled stop. As soon as we got there, we saw a Blackhawk medical evacuating a Marine whose hand had been blown off.

We came to a demobilized Iraqi semi-truck pulling a trailer carrying two huge missiles in the back. They weren't even strapped down; they had just been tossed on top of each other in the bed. We guessed these were Scud missiles but didn't know for sure. I pulled out a book I had found somewhere on Iraqi weaponry and located the picture and specs that resembled the missiles. They matched up to be

SCUD-B's that measured 35 ft. in length and nearly 7 ft. in diameter. They have the capability of carrying 250 kg to 500 kg of high explosives or chemicals.

Leaving the refuel site, we were bombarded with Iraqi civilians trying to sell or trade for cigarettes, soft drinks, turbans, or whatever else they could get their hands on. I bought a pack of cigarettes as a souvenir but I only had $10's and $20's and all they wanted was $1. Knowing how much it would probably mean to them, I paid $10 for one pack and got 250 Dinar in return as another souvenir.

We continued northward on our journey, passing many Iraqis waving and cheering us on before we pulled into BSA-6 at around 1230. We were 72 miles away from Baghdad. As we arrived, we could see two 7-tons full of orange mailbags! We couldn't wait ... it was like Christmas!

Chapter Fifteen

They called a formation around 1500, but we were there before they could get the words out! Words cannot describe just how exciting and exhilarating it was. I received five letters from three boys and two girls in the sixth grade at South Asheboro Middle School—the same middle school I had attended. I got a letter from someone named Laurie Sypole, a letter from Tammy's mom, Cindy, and a letter from my friend Morgan in California. I got a letter from Mom and another from Dad. And I got four letters from Abby. What a great day!

I read them in pretty much that same order, but I mixed Mom's, Dad's, and Abby's together because I wanted to hear from them about equally and I didn't want to read four in a row from Abby.

And I went through them trying to read as slowly as possible to savor every word written to me as if I had nothing else to do in the world. My eyes teared up through all of them, regardless of who they were from. I was just so happy. At least until I read the last one.

Friday, March 21, 2003

Eric—

First and foremost, I am mad at you! I have already asked you this question, but I am going to ask again hoping you will tell me the truth this time. Who is Morgan? And does she pose any threat to me? You have already told me both answers, but you must be leaving something out because she and her mother called your mother the other day. If she is no one special, other than a friend, why would she call your mother? The way it sounds you must have had something special with her in order for her mom to even know your mom's phone number. Do tell me and don't lie to me again! I don't like feeling like you are lying to me. But I don't know why in the world you would lie!?! I asked your mom who Morgan was, but she didn't know or she didn't want to tell me.

I'm sorry but I just know that sometimes you don't like admitting your past and you always try to protect me. So fuck both of those reasons, just tell me!

I'm a bit emotional right now so I'm feeling pretty insecure. I enjoy talking to your mom and I hate the fact that there is some girl in California that may feel the same way towards you that I do! That kills me! Well, I'm upset right now and I can't write anymore. I love you always.

Love, Abby

God—my mom has never even spoken to your mom!!!

I couldn't believe I just read that. Was there a joke in there I missed or something? I had to reread it three or four times because I couldn't believe she could have written it. This made me so mad that I started to get mad at her other letters, too—so she thinks this war is a mistake?

There I was, pacing. I needed someone to vent to. I had to tell Ryan and Russ about this. Maybe they would see something I was missing.

I was so mad I kept stuttering as I was trying to tell them about the letter. I had to just read it aloud to them. As I read, I would stop to curse at the letter. I was pointing to the letter and cursing like I was defending myself to the letter. Like the letter would get my point.

I needed to go to sleep. My day was ruined and maybe if I just went to sleep I'd wake up in a better mood like I always do.

Chapter Sixteen

The sleep didn't help. I woke up pissed. So pissed that when I was woken up in the middle of the night at 2200, I got up immediately without struggle because I was still so mad and in so much disbelief, I forgot to think about the bullshit of waking up five hours early.

I stayed awake all night repairing a power take-off unit, or PTO, on a Track. I was actually being productive this time, trying to keep my mind as preoccupied as possible by thinking about my work. After eating breakfast, I decided to express my feelings on paper in hopes of feeling better.

April 5

" First and foremost I'm mad at you!" ...

Who the fuck does she think she is? How does she get off accusing me of lying and proceed to cuss me out while I'm on the other side of the world at war? I guess it's justified in her mind because she doesn't approve of this war. Has she no knowledge of Saddam's past? Has she forgotten everything that happened on September 11th, 2001? How can she blame past Presidents for his actions?

I guess you can't get the full effect until you're standing over the burning body of a young child lying on top of other burning bodies of Republican Guard soldiers. Saddam had used the boy as a shield to protect his men from being blown up. Or maybe he trained him as a combatant. What a great fucking plan as they all lay smoldering on the side of the road after being blown out the side of the demolished civilian bus.

Or then there's the woman that was only identifiable as a woman through one exposed breast. This is the only part of her that remained as she was blown to pieces and burnt to a crisp. The rest of the former occupants lay in various positions and no one could tell which limbs belonged to different torsos. The only survivor happened to be the driver who was lying beside the bus begging for me to shoot him. Can you imagine seeing eight or so of your passengers dead beside you in this display? Can you imagine knowing that you're going to die because you're bleeding to death and beyond treatment but you can't die fast enough?

These people died for their ruthless leader. Yeah, Saddam is not to blame. He didn't know that such a thing could, would, and did happen. He didn't know we would invade if he didn't leave within 48 hours like we said we would.

Four Marines are dead from my Battalion. One just had twins but he never got to see them. This is the same guy that sang "Just the Two of Us" with his now fatherless son at a Marine Corps Christmas Party. But Saddam is not to blame, it's our past Presidents. Well, fuck all that bullshit! It pains me to even think about it, much less blame our own for it.

Who the fuck does she think she is? Who the fuck does she think I am? My love for her and her role in my life is nearly paralleled by the love I have for my Mother. I have never felt this strong for anyone in my life. I never thought that anyone could measure up to my Mom. But now this bullshit? It makes me have second thoughts.

I've been out here for 19 days without a shower. I'm lucky to get two MREs a day because we're on a food supply shortage. I have less than two MREs remaining which gets me by until tomorrow and then we're out. For the first time in my life I'm a day away from going without food. Our water supply is gone. I have to drink stagnant water from a puddle after adding Iodine tablets to hopefully kill the diseases. The only thing that I have to look forward to on a daily basis is for the sun to go down as it brings us a day closer to be heading home. It gives me a chance to miss Abby before going to sleep if there's no work to do. I talk to a star at night that I've named Star and ask him to tell Abby that I'm okay because I know she can see that same Star. And at a time when my spirits and morale soar to an all time high for the sole reason that I received mail, I saved the best for last only to be cussed out. The only sentence in the entire letter that wasn't bad was, 'But I love you always.'

What kind of fucked up ass shit is that? It sounds more like a chore than an emotional feeling. Did Spring Break, her Caribbean Cruise, lessen her level of maturity?

I mean come on Abby, let's be rational about this. You say that you're scared right now because I'm in danger. Does it really make sense to cuss me out? I'm sure the letter will really keep my head in the game. Are you scared for me being in danger or are you scared for you for me being in danger? Is this just a method to lighten your fears because the burden is too heavy upon your shoulders? I'm really at a loss—I mean she even said it in her letter. If I've already answered both questions then why would she bring it up again? Can I not have friends if they're not approved? Thinking back now, she had a problem with my friendship with Aaron who has been my best friend since the seventh grade.

How does she think she has the right to disapprove of Morgan? I met her for maybe an hour one day snowboarding with a guy named Matt and the three of us got to board together all day. I kept in contact with her and Matt because they're the only people I know in Cali who board and are fun to be around and hang out with. Well, except for Joe Diebert from Madera but I have no way of contacting him which is all the more reason for me to stay in contact. Morgan even helped me through my problems with Tammy as she was the only person I felt comfortable talking to. By the same token, I helped her out with her family problems. And when I tell her about Abby, she says that Abby must be a great girl and that she is happy for me. She had given Abby praise without even knowing her and giving no more explanation than "she is my girlfriend." And yet Abby freaks out because she and her mom called my Mom? I cannot help the fact that they called her nor would I want to. The fact that a stranger was concerned enough to call my Mom in her time of need, offering to help out in any way, gives me great pride and joy. Not only that, I have never given Morgan my Mom's number or even spoken to her mom. They must have looked it up on the internet to find it. But for Abby to freak out and accuse me of lying is nothing other than childish and uncalled for, period.

Oh, and to bring the mothers into this? That's taking it to a whole new level. How did Abby feel that Tammy's mom called my Mom? Her last line doesn't even make sense. Does she know that Aaron's mom and my Mom have never spoken? I've been friends with him a lot longer than anyone, so the fact that our mothers have never spoken is irrelevant. And then she says that my Mom didn't know or didn't want to tell her who Morgan was. Don't accuse my Mom of withholding information when she is telling you that others are concerned in order to keep both your spirits high.

The crazy thing is that Abby knows all there is to know about my and Morgan's relationship but just refuses to believe it. And then she accuses me of not admitting my past? I have no regrets about my past and I wouldn't change anything about it. Regret is a waste of time as things happen the way they are to happen and that is the only way they could have happened. If my past was changed, then I wouldn't be me for it is my past that has made me who I am today. My past may be made up of many different pieces that may be revealed at different times like a puzzle. The pieces are not

interchangeable and belong only to me. Without them, my life would
be incomplete; but with them, I am me.
Why am I putting up with this? I'm going to sleep!

This was a real roller-coaster ride of an expression of thoughts. I
would get so mad that I had to stop in the middle of writing and smoke
a cigarette to calm down. I would think I was finished at different
points, and then something else would get to me and I'd get pissed off
all over again. Finally I had nothing more to say and had reached the
point of questioning if I still wanted to be with her. But I wanted to
sleep on that thought; I knew I was really, really tired.

After a four-hour nap, I woke up in a better mood. For the
remainder of the day I worked to remove torsion tubes, support
assemblies, and road wheels from two hogs. I had post that night, and
because of my long nap, stayed awake until then.

I still couldn't get that letter out of my head, though. I felt better
after venting to myself but better is only better, not necessarily good.

After Nik relieved me at 2300, I went to my truck to sleep but
didn't feel like climbing up to my spot. So I crawled under the truck
and got comfortable but then got scared someone might move the truck
at night or in the morning while I slept. So I laid down beside the
truck—far enough away that I wouldn't get run over by it, yet close
enough that I wouldn't be run over by someone else's vehicle.

April 6. Daylight savings time began during the night, so I woke
up at 0430 instead of 0330. They said that daylight savings time
doesn't apply to Zulu time but they adjusted our Zulu time by an hour
anyway ... confusing.

I woke up aching; I had slept on my Flak Jacket over uneven
ground. I went to my truck to begin writing in my journal, but we
actually left on time. I couldn't believe we were actually ten minutes
ahead of schedule.

Originally, it was to be a twenty-mile march. About twenty-five
miles into it, we heard it was changed to fifty clicks. After the fifty
clicks, we heard one hundred clicks, and after one hundred, we heard
one hundred and eighty. Finally, after eighty-seven miles—one
hundred and forty-one clicks—we arrived. The odometer reading was
now at 497 miles; we were just 5 miles outside the center of Baghdad!

All of us had enjoyed the journey, and even though it had been exhausting it was actually quite memorable. We passed hundreds of supportive Iraqis lining the streets waving and cheering us on. The entire eighty-seven miles felt like a parade. We had to wonder if the war was already over.

We rolled through as Iraqis continued to come out of their homes and buildings to wave and cheer. Many of them had learned phrases in English and shouted, "Good, very good!" or "Thank you." Those that had not learned the phrase or were too timid to say it simply said thank you in their native tongue—"Shoo-krun."

We began tossing food out the window for them. As scarce as our food supply had been at times, there were still some things on hand that we just couldn't stand to eat. We tossed them rice, peanut butter, Charms and Skittles. Seeing their grateful reactions, we began tossing them food we liked. Knowing we were helping them was more rewarding than saving it for ourselves. And when that supply gave out, we opened the box of Humanitarian Rations. With so many recipients and only twelve rations, we opened each individual ration and began tossing out the contents piece by piece until all were gone.

We tried to aim for the children we passed. We knew this was an event they would never forget. They are Iraq's future; we wanted to make every effort to show them we are the good guys in spite of what they might have been raised to believe. The expressions on their faces were priceless. This was what we were here for. Occasionally we passed quarrels breaking out over our gifts, so we would toss more food at them to show there was no need to fight.

The festivities, as rewarding as they were, became exhausting. It had taken an entire day to make the trip. We were moving at about twenty miles per hour when moving, but stopped frequently. The heat was blazing and the air was still. Water couldn't cool us down, as our water supply was hotter than the air.

So when we arrived at BSA-7, I stripped down to my boxers trying to cool off, despite the gnats and mosquitoes feasting on me. We could hear yelling and shouting nearby, but weren't sure what was going on. It sounded like a celebration mixed with a religious ritual dance. We assumed it must be time to face Allah and pray or something. After all, we were sitting in Baghdad.

The wind was picking up as I got ready to go to sleep. I zipped my sleeping bag up only halfway because it was still so hot. A little later, I was woken by my sleeping bag slapping me in the face as the wind continued to strengthen. I zipped it up and was later woken again, this time by bombs over Baghdad. There was a lot of celebrating going on, yet there were also shots and explosions going off.

"Was it all over? Had we won the War?" I asked myself.

"Cock-a-Doodle-Doo!"

"Damn that rooster!"

April 7. I had been unable to go back to sleep. At 0300, we went to MOPP Level Zero; I had to put my cammies back on. When it was time to get up I was actually ready. I was bored lying there not being able to even close my eyes, much less sleep. But we didn't have any vehicles that needed to be worked on, so the day went by slowly. Everyone mostly just hung out in their little groups. I even volunteered to take some spare parts off a dead-lined vehicle that we had. Then I responded to some of my letters. As soon as the sun went down, I was asleep as the boredom had worn me out.

April 8. The next morning, I awoke and finished my responses to the letters. I wrote my grandparents. I wrote Mom and Dad to update them on our progress—I hadn't had a chance to write them since we left Kuwait. I mentioned Nasiriyah since I knew they must have been terrified by the news coverage …

> *"You may have seen us on the news. I've heard rumors that our convoy was shown on TV going though An Nasiriyah as "the convoy that went through hell," or something to that effect."*

But even as I was writing Mom and Dad, I still couldn't get Abby's letters off my mind.

> *" … Well, there you have the summary of our journey. When we finally received mail on the 4th, I got fifteen letters! It's really cool that everyone is so supportive, save Abby …"*

And then I wrote Abby a response just to see what I would write. I decided to wait to mail it until I received her next letter to find out if she apologized.

April 8

Abby,

"First and foremost," really? Are you fucking kidding me? How the fuck do you get off accusing me of lying after a good friend of mine calls my Mom to see if there is any help they can give? I don't understand you not mentioning the fact that I am at war and need all the support I can get. Almost 3 weeks of looking forward to getting mail and then you send me these accusations—this shit.

Oh, and this is after your disapproval of the war. I would have never expected anything like this to come from my girlfriend, you. I mean how do you think it makes me feel when at a point in my life I need you desperately, you come through on the negative side? Did you think I didn't have enough stress on my shoulders and needed more? Well good, you succeeded and gave more to me.

Do you know how I deal with stress? I will quickly tell you. When I feel stress coming on I get a "Fuck it, it doesn't matter," attitude—those are the exact words I tell myself. For example, in this case it would go—"Fuck you, you do not matter."

Usually the more I think about something, the more I can see the reasoning behind thoughts or decisions. In this case it was extremely difficult—maybe due to the circumstances. I don't like feeling like you are trying to control who I am friends with. I also don't like it when you cast down my friends when you don't even know them. These same friends give you praise, but yet you don't like them because you "know little about them." Think I am wrong? Think of Aaron.

I don't like the fact that you suddenly change from secure to insecure. I can only recall the days when I was insecure and my method of getting through it. Simplified, I mean that your security brought me security, but now your insecurity questions my security. Also, the fact that you would bring all three of our mothers into this bothers me. We have no control over them, so why argue over something that does not pertain to us?

You already know who Morgan is. So I am not even going to entertain the idea of defending myself or her. I shouldn't have to because I have never given you a reason to distrust me, and as my love and girlfriend, you should believe in me.

I am sending this letter because you do matter. I know that this war is stressful on everyone and that you just made a rash and unjust decision in your last letter. I am not going to allow myself to become a victim of this war and let our love slip away. Just as I have witnessed many inhumane acts since the war started, I will take it all in stride. Not to say that you are inhumane ... I'm not saying that at all. But that I will reason it with being just one of the many hardships of war. I will keep a happy heart and strong spirit about me, and together we will get through this. I think it would all be easier if I were there to discuss this. I can't wait to see you and I love you dearly.

Love always,
Eric

I never really made the decision to mail this letter to Abby, but it got mixed in with a stack of letters I had written to others. I mailed it without knowing it, and only realized what I had done when I looked for it a couple of days later.

Chapter Seventeen

At approximately 1200 on April 8, we were "Oscar Mike," or on the move again heading around Baghdad. We drove about 1.7 miles before pulling into BSA-8. Nik and I were tasked with building a bridge across a creek we were parked beside so we could set up our outer perimeter of security. We looked for some wood but couldn't find any. So we found the narrowest part of the creek and proceeded to knock down the embankment to create a land bridge across. We had a good time doing this as we got to play in the water.

A little later, I stumbled across the Gunner's laptop and a DVD. Moya and I watched *Enemy of the State*. A rooster was crowing again, but this time it didn't concern me so.

Then it began thundering and lightning, with occasional rain and a nice breeze. It was really nice lying down to sleep with the soft sound of light rain. The more I listened to it the happier I got and the more it seemed to cleanse my soul. Finally my face was wet enough; I zipped my sleeping bag up over my head and went to sleep.

April 9. I woke up the next morning at 0445, starving. I was out of food though, and went begging to Supply. It was a hard-fought battle but in the end, I came away with food for two more days.

As I ate my breakfast, I noticed a good-sized spider lurking beside me. It was just sitting there looking at me as if it were begging for some of my food or scavenging for what I might drop. I had never seen a spider like this. It was maybe three inches long by an inch and a half wide. It looked mean as it pointed two of its ten legs up at me as if it were putting up its dukes. The jaws on this thing were like the Jaws of Life and looked like they could snip a finger clean off.

Later I was informed it was a Camel Spider but that it had probably only been a baby. They grow to about eight inches in length and their venom is like an anesthesia. Supposedly they eat camels from the inside out; as their venom numbs an area they crawl inside to begin eating while the camel sleeps. How true that was I couldn't be sure.

Still with nothing to do, I opened my book to read and finished another two chapters. I also killed eight flies as I read. There were so many flies out here it was ridiculous. I had become very good at killing them, because the thought of them crawling on me had become so disgusting. Not only did they crawl on our own bodies and wastes as I've described, there were dead bodies around. I began to think about where they might have been before landing on my face. It was a horrifying thought, so I had to rebel.

I learned that if I tried to squash a fly by smashing it where it's perched, the fly would get away time after time, only to land on me once again. I knew there had to be a clever way to defeat it. What I tried next worked successfully about ninety percent of the time.

I would take both hands and place them about twelve inches from one another as if I was telling a fish story. Keeping them the same distance apart, I lowered them near the fly, keeping the fly in the very center. Once the knife part of my hand was hovering just above the surface on which the fly was perched, I began to bring my hands together slowly. The closer I could get to the fly, the better chance I had to kill him. The fly, sensing danger, would often sit perfectly still waiting to escape. When I had brought my hands as close together to the fly as I could stand before feeling it would fly away, I clapped them together. Dead fly ... victory is mine!

At 1430, we got mail again. I got mail from Carol and Dink Routh, Sarah, Dad and Abby. Dad's letter again made me smile. He had been numbering his letters ...

Thursday night
March 27, 2003

Dear Eric,
    I think this is letter #13, but I'm numbering it "14" for obvious reasons ...

Abby's envelope had two letters in it; they were dated March 26 and March 28. I was happy to get mail from her again, because I knew this would be an apology for her cussing me out like she did.

Wednesday, March 26
Eric,
    Hello, honey. Last Friday I received four letters but have not received any since. I get so excited to go to the mailbox, but I am disappointed when I do not see a red, white and blue envelope ...
    ... Last night I had a horrible nightmare. We were on our honeymoon—we went on a cruise. We were at dinner and you decided to go back to our room. Well, I wasn't finished so I stayed at dinner. When I came to our room I found you in bed with Ashley Love—I was devastated so I told you it was over and I walked out.

*That's when I woke up, and it felt like it all happened. I was so
scared I got up and turned my TV on and that's when I knew you
were at war—I was relieved you were there and had not cheated on
me ...*

> *Miss you more,*
> *Abby*

In her second letter she had written that she had just received two
of my letters and how excited she was. She was also sorry she had ever
questioned me about not calling from Germany, and for making it hard
for me when she had gone on spring break. Then she thanked me for
the advice I had given her concerning her mom and said her mom and
sister were actually there visiting, although from what she said the three
of them continued to have serious communication problems.

I guessed it was a good time for them to visit, being that her sister
and sister's boyfriend had just broken up. Was Abby just the next best
thing? It seems humorous to me that families become closer after a
breakup or loss of some sort. Does it really take all that? If only
families could remain close, regardless of significant event or loss. But
then maybe that just proves the power of family and unconditional
love. Family is always there, until they're gone.

She also said my Mom wants her to meet Kim. I wonder why? I
mean it's not like it matters to me but it just makes me wonder. Was
she doing it because she likes Abby and thinks of her as part of the
family? Was she trying to give me a hand because she knows how
strongly I feel about her? Was she taking Abby in because of the way
she talks about her own family? Or did it just come up in casual
conversation as a response to Abby mentioning that her mom and sister
want to meet Mom?

Abby wanted me to promise to tell her if she ever acted like her
mother. Don't worry, Abby—I'll be the first to tell you! Come to
think of it, I think Mom also made me promise the same thing.

Then she said that Aaron and Jess had broken up again. Didn't
this happen last year at spring break? Am I seeing a pattern here? Oh
well, I told him they should break up. I like Jess a lot but she's just not
right for Aaron.

Russ came by and wanted to know if Abby had written anything
about Kate. She really didn't have anything good to say about Kate's
and Russ's relationship. I didn't want to tell him, but at the same time I

didn't want him not knowing. I asked myself if I would want to know if the roles were reversed. So I let him read what she had written …

"I haven't talked to Kate. We never did go to dinner. But I did write her the other day (at boot camp). I'm trying to get brave enough to ask her what happened that night. She doesn't talk about Russ all that much. Every time I talk about you, I smile at just the thought of you. She doesn't do that, but that doesn't mean she doesn't care. I don't know—only time will tell, and hopefully the more I write her the more I will learn. I feel like I should write because she wants me to. I don't really know much about Kate but I think she is a nice person. But I also feel like she has ulterior motives. I just always felt like she was in love with you—that's what we had in common you know? Maybe I'm wrong. Russ is the only person who can decide that. I think she was in love with you and Russ treated her the way she wanted you to treat her—he was her rebound. Just a guess … "

I could tell it had upset him although he tried to not let it show.

At the end of her letter, Abby told me how Zula would smell the scent on my letters. It's kind of funny because like Zula, I smell the letters I get. It's amazing how scents will take you back to a certain time or place.

Abby's letters smell like her most of the time, but sometimes her scent is covered up by another scent that throws me off. Maybe she had written it after work and she smells like a bar? I won't ask Ryan or Russ that question because I know what they'll say.

Mom hasn't sent many letters, but the ones she has generally smell like her. She puts other things in the envelopes like race result print-outs and newspaper clippings which seem to neutralize the scent. Sometimes she puts Big Red in the envelope that overpowers everything. I wonder why she doesn't write that often.

Dad has a variety of scents. When he smells like himself it takes me back to riding with him on the way to school when I was younger. Sometimes he smells like Mom, and it takes me back to eating supper with the family. And sometimes he smells like resin.

Dad works now as an Operations Engineer for Concept Plastics, Inc., where they make furniture parts and decorative accessories out of resin. My Mom also used to work there as corporate secretary, and when I had been to the facility with her, I became familiar with the unmistakable smell of resin.

Sometimes Dad's letters had a strong resin scent, and sometimes it was weak. Oddly enough, both scents take me back to a specific location of the building. When it's strong, it takes me back to the factory where my Dad worked, and I can see in my mind's eye the doors entering from the offices. I look to the right and see the doors open to the outside. Looking straight ahead at about eleven o'clock, I can see all the workers perfecting the pieces that have just come out of the molds. And looking immediately left, I can see workbenches lining the walls where they've staged random items.

When the smell is weak, I'm taken back to the product showroom in the office area where Mom worked. I'm walking through the showroom from the lobby side to the breakroom area and the showroom lights are off. For some reason the lights were off every time I was there, but I could see my way because the doors were open on either side.

Ryan's girlfriend, Bridget, always sprays her letters with the perfume she wears. Reading her letters made it seem as if she was right there with you. Ryan would take his letters around and show them off to friends because of how good they smelled.

I had gotten two letters from Abby but no apology. I was having a hard time understanding. How does she not realize what she has done? Is she so mentally unstable that she isn't thinking clearly?

Later on, I lay down in my sleeping bag and Star is shining. I hadn't spoken to him since I received the letter from Abby so I told him I had nothing to say to him. Then I realized it wasn't his fault, and explained all that had happened. After talking for a while, I grew cold, wrapped myself up and drifted off to sleep.

Chapter Eighteen

April 10. I woke this morning and did not want to get out of my sleeping bag; it was freezing outside. But the word was that we were supposed to move out of BSA-8 by 0400. Well, that quickly changed to 0600. Then at 0600, it changed to 0630.

At 0630, we moved out like bats out of hell. Many Iraqis cheered us on and waved to us as we passed. The entire forty-mile march through and around Baghdad ended up taking seven hours and was nothing less than a parade. The Iraqis love us and think of us as their heroes. We all believed they were happy because they knew the Saddam regime was now over and their future was in the right hands. We couldn't have felt more proud to have been a part of it.

One man exclaimed in English, "Yes Mr. Bush; No Saddam Hussein, No!" Many could again be heard saying "Thank You!" or "Very Good!" Some took a more personal note with things like, "Hello. What is your name?"

This went on for the entire seven-hour trip. My arm had gotten heavy from waving so much. My face grew sore from smiling.

We arrived at BSA-9 after it had gotten dark. The odometer read 538 miles since we had first departed. Exhausted, I fell asleep. But as I tried to sleep, I could still see the children's faces as they waved at me. I actually woke myself up several times trying to wave back at them. I couldn't believe I was doing this.

I've done this sort of thing in the past on a few specific occasions. I remember going to the beach and body surfing the waves. And then at night, I would dream that I was catching a wave and would wake myself up, swimming away. I had also done something similar after a day skiing in the mountains. That night, I would dream I was skiing down the mountain and would wake myself up falling. It's a good thing I'm a better snowboarder than a skier.

April 11: This morning, I woke up feeling differently. The parade yesterday felt like a goodbye parade. It seemed like our job was over and it was time to go home.

I decided to quit smoking so I wouldn't come home a smoker. The rest of the day I spent writing letters and reading in my book.

April 12. This day was equally boring. Nik and I talked for a while. It seemed like all of our friends back home were at spring break while we were here.

So we came up with an idea for a screen-printed t-shirt. We were talking about all the fun stuff to do and the fun places to go for spring break. There are always people who buy keepsakes in memory of their good times and seem to love wearing t-shirts that are printed with the year, place, and event. We thought about all the "Senor Frogs" and "Fat Tuesday" shirts we had seen.

We decided we would make our own commemorative t-shirt. "2003 Spring Break" would be printed at the top; in the middle would be an outline map of Iraq with Baghdad magnified in the center. In the magnification would be the head of Saddam between crosshairs like he's in our scope. Around the map and targeted toward him would be renderings of our assault equipment—helicopters and jets flying overhead firing weapons down at him, ships in the Gulf firing missiles at him, and our AAVs, tanks, LAVs, HMMWVs, and troops below aiming in at him. And of course at the bottom of all this it would read, "Baghdad." We then decided that we could put the flags of all the countries and emblems of all the forces and units involved across the very bottom.

I began to sketch what I wanted the picture to look like. I got the shape of the country drawn and the bullseye over Baghdad before I realized I didn't know how to draw and this would be as far as I got. Maybe when I get back I can pay someone to draw it for me.

I opened my book to read and was able to finish Book 3. I decided that the tobacco I dipped while reading would be my last. And then I began finishing up some letters I had already begun.

April 12

Abby,

I got your letters from the 26th and the 28th. I was expecting one of them to be an apology but I was wrong. Maybe you didn't realize that you owed me an apology? Maybe there is a letter in between that I did not receive?

I can hear it in your writing. You don't sound like yourself. What is it that is weighing your mind down so? Is it me? School? Work? Family? Something is deeply affecting you and I want to know what it is. You told me about your nightmare and then you wake up and realize that I am at war—and you were relieved I was here? Come on, Abby, I would never cheat on you, but how could you be relieved that I am at war? You're not

*thinking clearly. Tell me what is clouding your mind. Is it the
war? Morgan?*

If it helps to put your mind at ease, here is the scoop
on Morgan. I met her over Thanksgiving of 2000,
snowboarding at Big Bear, CA. I was of course dating
Tammy at the time. We exchanged e-mail addresses and
instant messenger screen names. I didn't remember either
but about a month later, she IM'd me. We typed to each
other on a regular basis but there were never any phone
calls. Then, I guess it was early summer of 2001, she wasn't
on line often so we exchanged numbers and we began to
talk on the phone. At the time, and actually quite often,
Tammy and I had problems. Morgan was someone whom I
felt comfortable talking to so I would talk to her about my
problems. Also, the fact that she is 3 hours behind meant
she was usually still awake after Tammy and I fought.

We became really close to one another, but in a
brother-sister type of friendship. She invited me to come
out and visit so I made plans to take leave after New Years
2002. I flew out there for 11 days, taking my snowboard
with plans of doing a lot of boarding. Supposedly there
was a death or sickness in her family, and she ended up in
Virginia while I arrived in California. All I had was Morgan's
cell and I couldn't get hold of anyone there. I called Jay
and stayed with him for my vacation. I was so pissed, my
mind was made up to never talk to her again. She called
once she returned to Cali but at the time, I was in Vegas. I
cursed her and that was that.

A month or so later, she sent an e-mail to me
apologizing but understanding the fact that I never
wanted to speak to her again. I replied, telling her to call
me to apologize, and she did. I told her that it was her turn
to visit now because I was definitely not chancing visiting
her again. We continued to talk weekly after that.
Charlotte was officially my second home after Tammy and
I broke up, and Morgan and I talked every Sunday night as
I drove back to base.

In late summer, we made plans for her to come visit NC
in a December or January time frame. In late October, we
(Marines) got a brief saying that we were going to the
Middle East just after Thanksgiving. I told her to cancel her
plans because I wasn't going to be there. Then our date
kept being pushed back so she could have come but
something was different. I had met and fallen in love with

you, and I wasn't about to do something to interrupt what
we have together. I didn't have the heart to tell Morgan
that I didn't want her to come out because I had found
you. I took the easier of the two paths and lied to her,
saying all leave had been cancelled and we were on a 96-
hour notice to shipping off.

Once over here, it was a lot easier. I told her in a
letter all about you and how happy we are together. It
was kind of hard because it almost felt like a farewell letter.
I didn't expect a reply. She did reply, however, saying that
it didn't change a thing in our relationship. She couldn't
say enough about how happy she was for the both of us.
She said that you must be a great girl for me to talk about
you like I did and also that she was confident that you
would change my mind about girls after Tammy.

Concerning your questions: I have never been to her
house, never met her parents, never spoken to her parents,
and never given Morgan my parents' phone number.
Morgan is a good friend of mine and does not pose a
threat to you; if she did, I wouldn't be with you because I
would be undecided on who I want to be with. If I ever
became undecided for some strange reason, I would
break up with you. I would rather be single and free rather
than waste my time and yours through indecisiveness. I
would hope that you would treat me with the same
respect.

On a different note, although you are against this war,
it has been a good thing. And if we did train him years
ago, we cannot be held accountable for his acts of
violence and terror. Millions of Iraqi people have been
liberated by us. They come out of their homes to thank us
for what we have done. They admire us and are happy
that their future is finally in the right hands. I have spoken
with EPWs that only fought because Saddam had their
wives and families held hostage. They said that the people
who chose not to fight had their families executed. They
were stuck because they would be killed if they resisted us
but their families would be killed if they didn't. This war has
changed and will continue to change the lives of millions
of people for the better. I am proud to be part of this
humanitarian effort and believe that you can be
persuaded to support it as well.

Now to your second letter on the 28th. You sound a lot better in that one. I'm glad you understand about spring break.

Bad news—the pillow has a hole in it and I can't fix it. It was great while it lasted though, and if you read my journal you'll find some stories pertaining to it.

Sorry to hear about your sister's breakup. But beyond that, I'm sorry they used you as an escape. And don't worry, I will never let you become like them. That's cool that you met my sister. You two remind me of each other; you should get along great.

Send my regards to Aaron; I know he really loved Jess a lot. He could just never get her to treat him like he wanted.

You have good assumptions of Kate, and I think you are right. Russ, on the other hand, is shutting me out because he seems to feel he is in constant competition with me.

It's sweet that Zula smelled my letter, but it kind of makes me feel like a dog. Not because my letters are being sniffed by one, but because I smell your letters when I get them and can smell your scent on them. You smell so good and it smells like home. I miss you so much. I feel bad because you thrive on getting mail from me but I know there has been a huge gap in mail since my last letter before the war started. I will mail these the first chance I get, and hopefully that chance will come soon. I love you and cannot wait until this is all over and I am safe at home with you in my arms. Don't lose hope ... it won't be long now.

Love always,
Eric

At 1600, I had post with Nik and Velasquez. We told V he could go ahead and sleep because we weren't tired and were going to stay up talking for a while. At 2300, we woke him and finally went to sleep ourselves.

April 13. At 0700, we were woken, gathered all our belongings and went back to the trucks. I quickly realized my M&Ms were gone. There was a thief in our midst, and I was pissed. This type of thing normally wouldn't have bothered me so, but I was on Day 2 of not smoking and the first day of not dipping. I needed something to fixate myself on so I grabbed some cheese and crackers and got through it.

Later on, everyone got Oreo Cookies. Apparently Nabisco had sent our military forces care packages full of bite-size Oreos. We were all used to eating the same old stuff from our MREs, so these new Oreos tasted amazing!

We didn't have any milk to dip them in, but it didn't matter because we had water. I managed to get three packs, so I got creative. I made cocoa with half the amount of water I normally used in order to get a thicker consistency than usual. Then I dunked my Oreos and truly believed it was better than with milk (although it may have just seemed that way since we'd gone without for so long). More than satisfied after this delicious treat, I laid down for a nap.

SSgt Benson came over and woke me up to help him organize and move all our spare parts in the bed of the LVS. When they start riding my ass, I think it's because they miss me. So afterwards, I stayed at M-03 and ate supper with them. I thought this to be in my best interest because of the favoritism they were showing Russ and Ryan. If I ate with them and sucked up to them a little by showing some interest in what they had to say, maybe they'd get off my back.

As I lay in my sleeping bag before going to sleep, thoughts of my future came over me. What would I do if Abby moved to Wilmington? Maybe I could get a job at the local Yamaha and Sea-Doo dealership, Britt Motorsports, as a salesman or mechanic? What about a second job bartending? I'd heard of the Bartending Institute in Jacksonville but never gave it much thought. I think I'll look into it when I return home.

I could work Tuesday through Saturday at Britt's. And work Wednesday through Saturday at the club. I could more than double the money I'm making now and actually enjoy what I'm doing.

Then it hit me that I forgot about school. Maybe I cut out the dealership so I can go to class during the day. But I still want to do more.

What about writing? I've always admired writers although I've always hated to read. I think of writers as very distinguished individuals. I like to write and I could do it in my spare time. I quickly came up with a few book ideas … this war, love, motocross.

Then the idea came to me to write magazine articles. I've always liked to read magazines for new product comparisons. Maybe one day I could have my own magazine or even a publishing company?

So it seemed I might have a path to follow ... go to school as a business major with bartending, dealership, and writing options.

But what about staying active in the Marine Corps Reserves?
One weekend a month and two weeks a year could be a fall-back plan. This would also allow me to keep some benefits—both medically and financially. But with bartending on Friday and Saturday nights, it could prove to be a burden. Should I cut all ties with the Marines? I hate to burn bridges, but I know the Marine Corps well. Would the Marine Corps get in the way of my future aspirations?

I had a lot to think about, but for now at least I had some ideas. With all the possibilities, my mind raced to and from and it was hard to go to sleep.

April 14. I woke up this morning to find we would be receiving mail shortly. I got a postcard from Abby. What a great way to start the day ...

March 26
Florida Keys

Eric—
But Baby, What's up? Well, this was
our first stop on the cruise. We had a
great time but we got far too drunk at
Fat Tuesdays. We came back to the
ship and passed out while laying out.
I passed out and Brandy got lost—she
was so drunk she forgot where we
were laying at. I thought it was
funny, she wasn't amused. I love you
and miss you—wish you were here.

Love Abby

From Mom, I finally learned what the circumstances had been the day Morgan and her mom had called ...

My sweet Eric,

As I begin writing, it's Friday, March 21, at about Noon. On the news, they are announcing air raid sirens being heard in Baghdad—and the commentators are
XXXXXXXXXX

Scratch that. It's now Saturday afternoon, and I again am pretty much at a loss for words. Please forgive me if the rest of this rambles ...

Jimmie came in at the point I stopped writing on Friday. He had brought me lunch—I have not wanted to leave the house (or the phone) for the last few days. Just as he walked in, news coverage showed all hell breaking loose in Baghdad. They had already announced the loss of 8 British and 4 American Marines in a helicopter crash, and one other Marine officer. But then the news of the loss of another Marine was announced, with no other "descriptive" information. I tried not to even think about the "what ifs" until Jimmie left to go back to work. But then, as much as I tried to not worry, it got hold of me. I kept thinking about what it must be like for a family to get that kind of news—thinking it just could not be you and yet still wondering.

About an hour passed—still with no more information about the Marine who was lost—and the phone rang. I was afraid to answer it, but I did, and some unknown female voice on the other end asked if I was Mrs. Cox. I could tell she was not a telemarketer. She began identifying herself, but I was confused. I could hardly speak, and felt absolutely paralyzed until I finally heard her say something about "Morgan." She was Morgan's mom! She wanted us to know how much she was thinking about you, and us, and how much you meant to Morgan. I tried to explain to her that I was having a particularly difficult time at that moment, and wasn't always in such an emotional state—then she put Morgan on the phone and we talked (and cried) for a while. They were both so sweet, and will never know how much I appreciated their call.

Then, a few minutes later, the phone rang again—it was Abby! Unbelievably, she was in High Point on her way to see me. In about five minutes, she and Zula showed up. She brought me a beautiful little box she had bought for me on her cruise, and an adorable card. She, Zula and I sat out on the patio and talked for a couple of hours (about how wonderful you are, mostly), and then Jimmie came home and we went to the Liberty Brewery for dinner (the place you went once in High Point and liked).

I am continually surprised by her thoughtfulness, and sometimes even feel guilty in view of how I treat Jimmie's parents. She is having such a hard time—besides not having you here, her situation at school, being sick, and even having her mom and sister coming in today. I told her about the time when you went to basic, Jimmie lost his job, then the wreck with your truck, etc. ... when I thought I was going to lose it ... then got your letter from basic saying the wrecked truck and lost bikes almost seemed funny to you—and your letter changed my entire outlook. Once again, we need some of your sense of humor—but I imagine that's even hard for you to come by right now. After she left, I felt a little better and actually slept for a few hours.

We got your post card yesterday, and the letters for Kim and David. I called Kim to tell her I was going to bring her letter to her, but she was so excited she insisted I open it and read it to her on the phone ... something I will not do again. It's apparent you have tried to keep your letters to Jimmie and me much more positive than you really felt. It breaks my heart to know how lonely and isolated you must have been. I can only hope that now, with the war finally being fought, you are simply too busy to have time for thought, and that after the fighting is over, things will be organized in such a way that will let you all get your mail and packages regularly.

I went through some of the letters we had written back and forth from Parris Island a little while ago. I found one I had written you, and wanted you to read it again. God, I wish I had the words to comfort you and somehow make things better like you could always do for me. I can only assure you how very much we love you, and how much Abby loves you. Life goes on here, day by day, but it will never be okay again until you come home.

All my love,
Mom

Saturday, Oct. 2 – 5:00 PM
(1999)

To my Marine-in-the-Making:

So sorry I haven't written for the past couple of nights. I had promised myself I would write you every day, but keeping both that promise and the other one I made to myself that I wouldn't write you a letter when I was having a really bad time without you just wasn't possible. Certainly don't want my letters to be a downer in any way. But today has been a little better—not that I miss you any less but at least I have been able to change my thoughts a little.

I'm trying to stay as busy as I can on the weekends to make the time pass faster. Today they were having an antique fair in Cameron—just a few miles from Hang Time. You can imagine the memories that came back along the ride—from your very first race at Devil's Ridge to the mid-summer practices at Hang Time when I felt I was going to die from the heat just watching. I remember Jimmy and Linda at Hang Time, and how much they liked you and Brandon. I remember the time Brandon got so sick from the heat and had to stop, but you just kept on riding in spite of it. I remember how much you wanted to win those first trophies at Hang Time, and how none of us ever had any idea of just how many trophies you would end up winning over the next three years. Nothing ever stopped you—not heat, cold, or rain—(except that one day in the mud at Devil's Ridge). Remember the time at Cathey's Creek when you slept in the trailer with the tarp over you, and I slept in the van? The next day it rained—I remember the mud at the starting gate and sliding down the hill—and yet you kept going and still seemed to love it. Remember Turkey Creek, and how mad Jimmie got trying to maneuver the motor home and the men with the dogs.

Today I drove through Carthage—and passed Hardee's and the antique shop where you and I stopped. All the memories. The trips you and I made together before Jimmie got involved. The times I saw you help another rider even if it meant coming in last. The times your attitude and good sportsmanship put my anger to shame when I thought another rider had treated you unfairly. Your courage and determination to win, but to win fairly. I cherish those memories, Eric. I cherish the life you gave me. I cherish you! Your school years—the day I got a call at work from Charlene Smith at Teachey (4th grade?) when she was so upset because apparently one of the girls in your class was leading a discussion on "blow jobs" and you were a part of it. She was so alarmed—and it was all I could do to keep from cracking up on the other end. I had to pretend how concerned I was, and assured here I would have a serious talk with you. The panic I felt the time I came home and found the note on the door from Mrs. Raynor saying she was so sorry about your accident and was there anything she could do—and I didn't know what had happened or where you were (bicycle accident). The time I got the call saying you had been hurt and were going to the hospital, when you had been riding the skateboard down Grandma's drive, rolling over the grass to a stick, and diving into the pond. To say nothing of the time I got the call from you, and then Grandma, when you and Julie were having your little party. (I never did know exactly what happened _before_ Grandma got there!!) I think that one episode impressed Paul with you as much as any other I ever told him about. He found it pretty impressive that you would get into such a situation at 11!

The time when in your own special way you quit Alex's baseball team. (Somehow I'll bet there have already been some instances with your DI's that you have wished you could do the same.) The infamous Hang Time crash—and all the other spectacular crashes you managed to have. And then most recently, your little party in April when we were gone to New Orleans. Although I couldn't believe you at first, when you finally convinced me you were serious, it just seemed like something you would do. (I think I'm going to go ahead and tell Jimmie about that soon. He's missing you so bad he won't get upset.)

No fear. No fear of me, your teachers, the track, or whatever else you had to face. Just like with the party, I guess your joining the Marines is something I should have known you might do. One of the last things I would ever have chosen for you--but then neither was motorcross. Seeing you ride--watching you get better and better, seeing your self-confidence grow—brought me more happiness than you can ever know, while at the same time, more fear. In a strange way, I wonder if that will happen all over again with you in the Marines. It just hurts so much not to be able to be there this time. It's like the dream you told be about when you crashed on your bike and couldn't find me—you have no idea how much that dream has haunted me since you've been gone. Yes, I know—you were a whole lot younger then and your dreams certainly aren't about me any more--but still, I can't help think about it. I hope you understand.

I'm going to close now—it's Saturday afternoon and once again I feel so much better knowing tomorrow you will have a little bit of a break in your schedule and can relax a little. If there is any way you could write—just a line or two to let us know how you're doing—please do. I don't think I'm going to be able to make it much longer without hearing from you!

69 DAYS AND COUNTING!!!

all my love,
     Mom

Mom didn't sound so great. I remember thinking in boot camp when she had written this letter that she sounded depressed. That doesn't even compare to the way she sounds now.

And as I was reading the letter she had sent me in boot camp, I realized she had called it correctly when she wondered in a strange way if her fear would grow. I understand now why she isn't writing me out here very often. I have to write her a letter to cheer her up ...

April 14

Mom,

You have got to cheer up! I'm fine—as always. I mean it's not like I have never had anyone shoot at me before. Those rednecks behind our house didn't like it when dirt bikers rode on their land. Then, I didn't have a gun to shoot back, plus I wasn't expecting it. Now I and everyone else around me have machine guns. Furthermore, we are the hunters—not the hunted.

So you need humor? Well, get this ... coming from our valiant leadership. In our convoys, I ride in the middle seat of a regular cab 7-ton truck. Kind of like riding in the middle of Dad's old Dodge. Well, we were briefed prior to going through An Nasiriyah that contact would be inevitable. They told us Marines had secured the western side so don't fire west (to our left). As we drove through, we quickly learned that nothing was secured as we received fire from both sides. I was shooting out the left (across the driver from the middle seat), and Sgt. Oyster was shooting out the right. Our convoy made it through unharmed, but that's not the funny part. Our Bn Maint Officer, CWO3 Perser, riding in a Hummer with Cpl Grimsley driving, jumped into the back seat, barricaded the doors with packs and whatever else he could find, grabbed Grimsley' rifle and held onto his own pistol, and ducked down on the western side—the side we weren't supposed to receive fire from. This is the man and leader that we all work for.

But that's not al ... it gets better. A couple of days later, he gives us a "motivational" talk. In this talk, he tells us that he has no fear. All of our mouths drop open. He says again that he fears nothing out here, and he would get out and walk down a dark alley and fear nothing, because he is the one to be feared. He told us we didn't need to think of him as a hero because that was just his mentality.

We can't tell him what we think of him now and just have to suck it all up. We've got to be more tactful. We've talked to our CO and there's a "gong show" being planned for his entertainment once this is all over ...

... I'm not mad at Abby any more. She is just in a bad mental state and probably doesn't even know what she is writing to me. Kind of reminds me of you right now. Both of you need to stop worrying so much and understand that I will be home soon. I'm happy and proud to be over here and I'll have a wallet full of money and a chest full of medals when I come home. Not to mention the stories, along with the simple fact that I have been part of history over here. And who knows, maybe this is what I needed to help me decide or at least give me choices on where I want my life to take me. My biggest fear has always been that I wouldn't succeed in life. I never thought I could measure up to Dad. But now for the first time in my life, I feel that success is near. It seems to just be a matter of time, rather than a question of yes or no ....

Love always,
Eric

Russ came over to visit with me after he had read his letters. Kate had sent him pictures and there was one of Abby at Mythos that he gave to me. This was just what I needed! It felt so good to look into her beautiful eyes.

Later on, I ran into LCpl Pez, who is a Reservist. We got to talking about the Reserves after my thoughts the night before, and he gave me hope about staying in. He says as an NCO there is no work to do and I would just have to keep up with the troops and make sure they're working. He said I would make about $100 a day, and that it's really easy to get out of work if there is a scheduling conflict.

But then I got to thinking about the picture that had been painted by my recruiter of my future in the Marines when I first enlisted, and how that had worked out for me. I figured I needed to do a little more investigating this time ...

April 15. I got up this morning and tried to finish some letters so I could mail them off. But before I could finish, I was informed I was to leave at 0500 to provide security for a convoy going south into

Baghdad. We were to get parts, mail, and food from our supply point, CSSB-10 (I don't have a clue as to what all that stands for).

The convoy was good for me, mainly because it gave me a change of scenery. But I soon realized my happiness and motivation levels had fallen. I was riding in the back of SSgt Thompson's AAV while standing on top of the seats with the cargo hatch open. It was just another parade through Baghdad with everyone waving at us. I still waved to the children; we are their heroes and it made their day to see us wave to them and I didn't want to let them down.

But I was growing tired of waving. I didn't want to smile any more. I'd seen enough of these people, and I want to go home and see my people. I'll be thrilled to smile and wave at people when I get home, but I'm finished here.

When I returned back to BSA-9, we handed out mail and I received a package from Abby. I opened it as fast as I could and found: *Sports Illustrated Swimsuit Edition*, thirteen packs of cigarettes, seven cans of dip, flashlight with batteries, Twix, Starburst, Big League Chew, Altoids, book of matches from Mythos with Kate's number inside, box of wooden matches with pin-up girl on front and only one match inside, and a phone card in Spanish.

I spent the rest of the day trying to put together the riddle that she apparently sent me. I mean, what was she trying to say here?

Maybe the one wooden match meant I'm the only match for her? But then it doesn't make sense why she would have sent me Kate's number and a Spanish phone card that I can't read.

Then I started thinking about thirteen packs of smokes and seven cans of snuff. The matchbook with Kate's number had twenty matches. The same total number of tobacco products. Why the magazine of half naked girls? Maybe she is asking me to ask myself a simple question while she answers it here?

If she has given me the one wooden match to prove that I am the only match for her, is she asking me if she is the only match for me? Has she given me past and future options and asked me to choose the one single match for the rest of my life?

This doesn't explain the thirteen and seven though. Am I missing something there? Do those numbers signify something about our relationship that I am not picking up?

Nothing seems to make sense. Maybe the matches and magazine were the riddle and the tobacco was unrelated?

Maybe it wasn't a riddle at all and I just have too much time on my hands? Maybe I'm just going crazy?

I tried to call home on the satellite phone later on, but there was a muddy storm about us and the reception was bad. Plus the battery was almost dead, so it didn't matter anyway.

The land here is so dry and dusty that all our vehicles and destruction have created quite the dust clouds. The storm clouds are rolling in, but they're a nasty reddish-orange color. Looking at them made me think of a bleeding country as the dirty water spilled down upon us.

Chapter Nineteen

April 16. It was so windy this morning I didn't want to get out of my sleeping bag and got up late. Ryan came by after getting off post at the ECP and had gotten me a turban. So I talked to him and Russ for a while and then went back to the truck because it was so cold. I wrote a few more letters because I found mail was going out, and then read more of my book.

Then Ryan came back by to get me to go make some phone calls. He tried to call Bridget but got her voicemail. So he called his mom and talked to her for a couple of minutes.

After he got off, I called Abby but for some reason knew she wouldn't be there to answer. I called her cell and her apartment, but she didn't answer either so I just left her a voicemail on her cell.

Then I called Mom and Dad. Mom answered and was so excited to hear from me that she was laughing and crying and trying to talk all at the same time, and I couldn't understand a word she was saying.

I had been worried that she would hang up on me when I placed the call because there was a five second delay from my end to hers. Five seconds doesn't seem like a huge delay, but it's more than enough to really mess things up when trying to have a normal conversation— especially when I was afraid I was going to get cut off anyway. If I waited until she answered and it was my turn to talk, then she would have to wait five seconds before hearing me and I knew she would mistake me for a telemarketer.

I knew how much she hates telemarketers. If she's in a good mood when they call, she just hangs up. If she's in a bad mood, she will stay on the phone with them and bless them out. It usually ends with them hanging up on her because she's asking for their home number so she can call them at supper and interrupt *their* family time.

So to avoid this confusion, I had to begin talking as soon as someone answered to reduce the initial delay. I also had to explain the delay in that initial speech so the conversation would run more smoothly. It actually works best if the phone is used like a 2-way radio and one person says "Over" when it's the other person's turn to talk.

After Mom finally calmed down from the initial shock of speaking to me, and with our time lapse problems, she told me the same thing she had told me in an earlier phone conversation—that she was going to quit trying to talk and just wanted me to tell her as much as I could about what was happening. Three minutes into the call we were disconnected so I called her back to continue. Three minutes into the second call we were disconnected again, so I decided to leave it at that. But my mission had been accomplished. All I had really wanted to do was let Mom and Dad know I was O.K. I knew the media coverage must have scared them. I knew they wouldn't receive any of my letters for another three weeks. I couldn't imagine how terrified they would be if they just stopped hearing from me after the war had commenced.

I don't know why, but I was really nervous when I was making the calls. I didn't have a clue as to what I would say. I didn't know how I could comfort them with such a short phone call. Maybe that's why? It felt strange.

Later on I got a package from Mom and a lot of mail. Mom's package had four magazines in it, along with food, Ranch dipping packs, playing cards, chew, and two pencils that didn't skip when I was writing in all the sand and dust. There was also a Hallmark Shoebox card in it that had an old lady on the front with big purple hair, purple eye shadow, and bright red lipstick. Her purple and blue outfit with brown accessories doesn't match and she's holding a green tote bag with a Chihuahua wearing a purple collar in it. The woman is saying, "We can LICK ANYTHING if we try!" And the dog is thinking, "How true …"

Abby mentioned in her last letter that she had sent me a package. She didn't mention she put together a riddle for me to solve, but she did say it should entertain me. I still hadn't been able to figure it out and it was driving me crazy!

Ryan came back by a while later; he was again able to get a hook-up with the phone watch. Somehow they had become friends, so we wouldn't have to worry about time limits or anything. We decided to stay up late so no one would be around to interrupt our phone calls to the girls. He, Russ and I talked for a while waiting for everyone to go to sleep. Then we made our way over but found a few people still there. We started talking to the phone watch so he would be cool with us using the phone for a while longer than we would have normally been allowed.

Ryan called Bridget first. They talked for about ten minutes or so while Russ and I talked to each other to give him some privacy. Then it was my turn.

I called Abby, and this time she answered. She acted the same way Mom did when I got her. I finally calmed her down and could tell she was crying. I assured her that I was O.K., and that I would be home soon. I felt less nervous about this conversation, and told her how much I missed her and she told me the same. I felt bad though because she was crying and I wasn't.

I thanked her for the package and had to ask her about the riddle. She apparently didn't even know what I was talking about. I reminded her that she sent me a book of matches with Kate's number, a Spanish calling card, and one wooden match. She replied it was totally coincidental, and the other matches must have just all fallen out.

I didn't know whether to believe her or not, but as we talked, I realized she was telling me the truth. Then she suggested I shoot myself in the foot so I could come home. I thought to myself, "No, I need that foot to cram up your ass for cussing me out while I'm at war," but I didn't say it. I told her she didn't have anything to worry about as far as Morgan or anyone else was concerned. I reiterated that she was the love of my life, and no one could ever replace her.

Abby and I talked for seventeen minutes. Afterward, Ryan, Russ and I used the phone to call a pizza place in New York where we proceeded to order enough pizza for our unit. We told them we were a military unit flying back from Iraq and would be there in two to three hours to pick it up. We told them none of us had a credit card number to give them because all of our personal effects were in the baggage compartment. I think they believed us for the most part, until we started giving them porn star names like Ben Dover and Mike Hunt and laughing.

I tried to stick around and hang out with everyone, but it was freezing cold. At 2100, I was cuddling with my sleeping bag and happy.

April 17. I went to check in at 0300 in the morning. After check-in, I went right back to my sleeping bag to take a nap. I hadn't had much sleep the night before because my mind kept racing with thoughts of home and Abby.

When I woke, I climbed into the truck and did some crossword puzzles from one of the books Mom had sent me. Then, when I looked up from the puzzle after a while, I couldn't believe what I saw!

There was a volleyball game going on in the middle of our BSA. A game of horseshoes was being played beside it. There were some Marines throwing a football to my right. Then there was another game of horseshoes. Chapman and Recon Ricky were both wearing John

Deere hats. Over to my left, some guys were playing softball. And beneath the camy netting, there were random games of spades.

"What the hell are we doing here?" I wondered. *"I want to go home! If the war is over, which it apparently is, then let's go home!"*

But then I realized what was going on was better than being bored and *not* playing.   I got out of the truck and began tossing football with SSgt Benson, Russ, and Ryan. Tossing a football is only fun for so long, so I tried to get a game going. But they didn't want to play so I left to go to the volleyball court.

Volleyball is really fun, although it took a while to get to play. All my friends were tossing football, so no one playing volleyball knew if I was any good. When I finally did get to play I proved myself and we were the winning team for about a dozen or so games. We finally tired and elected to give up the court.

I was starving anyway, and headed back to the truck to finish my breakfast from earlier. Then we got mail and gifts. The company gave us extra t-shirts and socks as an "at-a-boy" gift. (We'd been wearing the same stuff for nearly a month now, and it had gotten pretty rank. We had packed six pairs of socks and six t-shirts, and had worn each for a week before putting them in plastic bags. We were supposedly able to do laundry in a bucket but that had not worked out very well. The water would turn grey and milky like we were mixing concrete. Plus it reeked with odor and would attract flies. So instead I had elected to lay my socks on the rooftop of the quadcon at night to let them air out.)

I got a letter from Navy Federal Credit Union saying I had received a tax refund of $763. Then I got a package from Kim with a lot of goodies in it. My only problem with getting goodies was that I never knew how much I should eat. If I ate too much, then I might run out too quickly. If I ate too little and happened to get more in the near future, I had to give stuff away. And it's not that giving stuff away is a bad thing, but out here, everything carries a price tag of some sort. So it's good to have things of value to trade.

At about 1600, I went to sleep, wondering if the war was really over and about everything that must have happened. At 2300, I was woken by a firefight of some sort in the woods just about three hundred meters from our BSA. This had to be right in front of our security Tracks, so I waited to hear to see if they lit anyone up. Nothing happened, and I fell back asleep.

Chapter Twenty

April 18. This morning at check in, we all got a hygiene package and more socks. Shortly after that we got mail; the mail we received was over a month old.

I got a letter from Aaron, and one each from Mom and Dad. Both their letters had been written March 14, before the war had even started. On that date, they had just received the letter I had written them on February 22. Dad's letter was full of news about the family, but Mom had other things on her mind ...

My sweet Eric —
As I start this letter, I'm not sure I will finish it. It's like always—I can't stand to write you when I've had a particularly bad day because I want so much to be upbeat and positive.

We got your letter yesterday, the one written on Feb. 22 describing your trip to the port and what you saw along the way. You also told us about the "comforts" the other services had—phones, PX, and the most heartbreaking to me—hot meals. I just cannot understand how the Marines can be treated so differently, and why you cannot have those things. And to see things from your eyes—camels, shepherds, and the other things you told us about—it just makes me feel so very, very far away from you, and so very helpless. I know you miss Abby terribly, and how much it hurts. And then I hear news reports today about these six African members of the UN who want to keep us from going to war for another four to six weeks. It was at that point that I went screaming out of the room in an absolute tantrum. How in the world have we as a country gotten ourselves in this position, and how long are you and the others going to have to sit over there in tents in a desert waiting for our leaders to decide what to do? It seems insane. I just want you home.

... Eric, please know that not every day is like this. Some days are O.K. Like last week, when we were able to talk with you, and hear the smile in your voice. Maybe tomorrow will be better ...

... Take care of yourself, Eric, and know how proud we all are of you. We look so forward to hearing from you soon. It has been five weeks now since that morning you left Camp Lejeune, and by the time you get this letter, two or three more weeks will probably have passed. If the saying "nothing very good or very bad lasts very long" is true, this can't last much longer.

All my love,
Mom

I finished reading and took a nap—fly free! I laid one of the cloth rags Mom sent me in a package across my face. It was thin so it didn't bother my breathing or make me hot, and it was clean so it didn't attract flies.

I got up from my nap about 0730 and it was time for breakfast. Field mess had half-pint cartons of milk and small bowls of cereal for us. It was so good being able to eat something besides an MRE. The milk tasted amazing, even though it was warm. I dipped my last pack of Mini Oreos in the leftover milk until it was all gone.

I was thinking about doing a crossword puzzle and then maybe playing a game of volleyball. But mostly I hoped we would keep to the schedule of leaving out of this place at 0200 in the morning.

Today marked one month of being gone from Camp Matilda. It was also Good Friday. And it had been thirty-two days since my last shower.

Rumor had it that in five days we would be in Camp Fox, Kuwait. I can't wait to be out of this country. Strange as it might seem, Kuwait would be the good life right now. And hopefully we'll be flying back home sometime between May 28th and June 3rd.

Although I knew it wouldn't be smart to get my hopes up, I couldn't help get excited thinking about it. So I decided to go around and take some pictures of everyone here; after all, I had eight disposable cameras to get rid of. Then I got into a pretty competitive game of volleyball. I tore my rotator cuff trying to spike the ball. I continued to play for a while longer, but the pain reminded me that I needed to take it easy.

Back in '97, I was racing motocross at a track in Orange County, NC, near Raleigh. I was leading the race and got ballsy trying to clear a triple jump. The problem was that I bobbled on the double before it and realized I wasn't going to get enough drive to reach the speed needed to clear the triple. So I used a technique I had learned called "preloading." I slam into the face of the jump and collapse my body into the bike to fully compress the suspension. On liftoff, the suspension rebounds with greater force, sending me higher than normal and in this case, hopefully farther.

Unfortunately I came up short and cased the third jump of the triple. In other words, I landed on the up ramp instead of the downside that would have allowed me to smoothly transition into the following turn. My suspension bottomed out and rebounded abruptly, kind of resembling a bucking bull. My feet and legs went off the bike and I looked sort of like Superman in flight as I clung to the handlebars.

My right hand was still holding onto the throttle, twisting it wide open. The combination of the circus act, the acceleration of my bike and the subsequent crash that followed ripped my shoulder out of socket and tore my rotator cuff. I continued to race in the coming weeks, never allowing my shoulder to heal properly. So it's not unusual or uncommon to re-injure it.

At 1400, my shoulder was killing me and I was starving so I went to my truck to search for food. Ryan and Russ came by and had great news. They had learned that at our 1530 check-in, we were to be told that reveille was to be at 2330 tonight … we're to begin our retrograde back toward Kuwait at 0100!

We were so happy and having a great conversation. So many times during this adventure, with so many things on our minds, it had been hard to talk. This time, we were cutting up and snapping pictures of ourselves; there was no searching for words!

After the 1530 check-in I went to my spot to sleep, but of course was too excited to do so. I just lay there wide-awake with my mind in overdrive, unable to hold a thought for more than a couple of seconds. I decided I needed to sing myself a soft song to calm down. What song did I know? I like the "Somewhere, out there" song, but I couldn't remember the words to it. I just made up my own …

"Somewhere, out there; beneath the pale Star-light, Abby's thinking of me, and loving me tonight."

THREE

Chapter Twenty-One

RETROGRADE!!

At 2330, I woke but didn't get up until 0000. Then at 0100, April 19, we amazingly rolled out on time. It had taken us the whole war to get it right, but we did finally manage to do it.

Our convoy speed was about 25 miles per hour, not accounting for the frequent stops. It was dark as we rolled through the streets of Baghdad, alert for anything. There were looters out and we were ready in case they were more than just criminals.

We passed the Presidential Palace, and after about an hour, reached Highway 8. Apparently we were to travel that road for a while, but had to wait to be cleared to enter. We sat underneath the overpass and on a ramp next to rows and rows of houses. We were all concerned that this location was not the best spot for us to be. The buildings were in the dark, while our convoy was out in the open under the lights of a large street. We were once again sitting ducks for anyone who wanted to rebel, so we sat perfectly still, scanning the darkness, ready to fire.

After half an hour, though it felt much longer, we began moving. We drove south but it was going to be a long journey; our vehicles were on their last legs. Tracks were continually breaking down and had to be fixed. The ones that couldn't be fixed had to be towed. Towing vehicles is twice as hard on the vehicle doing the towing, so even the good vehicles began to go down.

Finally, at 1430, we pulled off the highway and stopped at an old building. This was to be BSA-10, 136 miles south of Baghdad. We hoped it would only be a short stay.

I had to go take a shit that I had been holding for a couple of days. I had to go so bad that I asked Russ to hold my rifle and took off. I didn't even grab a shovel because I wouldn't have had time to dig a hole. I hurried out the back of the building into a field overgrown with dead weeds. As I scanned the field in search of a hidden location where I could do my business, I began to worry about land mines and unexploded ordnance. I walked carefully.

I reached a trench of some sort and climbed down into it where no one could see me. This was the first crap I had taken where I'd had total privacy since we left Kuwait—or at least that's what I was thinking.

I was unaware that no more than twenty feet away from me lay a snake watching me. He was actually up on a ledge in the same trench I was in, watching me at eye level.

Thinking I was alone and not even considering I could be in danger, I took my time; I was enjoying the solitude, as it didn't happen very often. When I finished, I unknowingly walked in the snake's direction. Less than ten feet from its spot, I gasped and froze as I spotted him. He was poised, ready to strike.

This snake looked meaner than any snake I'd ever seen. It was black and appeared to be coated with extra slime; it was so shiny that it looked silver in spots. He was coiled up like a garden hose with his head draped over his body. He must have been a good four or five feet long and had a girth just larger than a soda can.

I've never been one to be afraid of much. Actually some would believe I'm close to fearless. But when it comes to snakes, I am terrified. I don't even understand why, and in the past had tried to overcome my paranoia. But I hadn't been successful.

We stared at each other for what seemed to be a couple of minutes as I tried to slow my heartbeat and catch my breath. I decided I would make one more attempt to overcome my fear.

Why did I leave my damn rifle with Russ? Why couldn't I have grabbed my E-tool? I'm going to walk closer to it to see what it will do. I have to prove to myself there is nothing to be afraid of and that I'm faster and more terrifying to him than he is to me.

Just as I was about to take my first step, a fly buzzed by me. My paranoia took over. I knew the fly buzzing by was actually the snake lashing out at me. I leapt back like a scaredy cat, but it didn't stop there. As I ran back through the overgrown field towards our BSA, every tall weed that brushed against my leg felt like the snake raring up to bite me. I began hopping sporadically and every time I came down to earth I thought I could see the snake underneath me. At full speed, I jumped and ran and jumped and ran until I reached the concrete slab of the building and got some relief.

I was completely shaken and a little worried that I might get made fun of, but I still had to tell someone. Well, "someone" quickly turned into a group who listened to my tale of the encounter. Many were frightened by the story, and others talked of going to hunt the creature down.

Someone had a book of venomous snakes and spiders, and I began to try to identify it. When I saw one picture I recognized it immediately. It was a Desert Cobra and I was terrified again. And no one went to hunt it.

When I had calmed down a little, I couldn't help but think about what the snake was thinking after seeing me running from a fly. I pictured it like a cartoon, with the snake just laying there in the same spot, giggling a sly laugh and calling me a coward.

Shortly afterwards it grew dark and was time for sleep. But it was lonely. We were under shelter now, and there were no stars to look at or talk to. And tomorrow would be Easter, but there wasn't anything to look forward to.

Sunday, April 20. It wasn't a happy Easter. There were no loved ones to go to church with. There was no lunch at Grandma's house. And there was no Easter Egg Hunt for all the little ones in the coming generation.

Early in the morning, I helped SSgt Benson inventory all his parts in the containers on my truck. During this, Gy Foust decided to switch SSgt Benson and me so he could be on the truck where his parts were. So I gathered up all my belongings and moved out to the R-7, M-03.

It was actually nice being able to leave the truck I had been riding in and move to the vehicle I had worked on daily since arriving at Camp Lejeune more than two years ago. Plus, I was back with Ryan, although SSgt Lopez was on the Track, too. His first words to me were, "If you're going to be on this vehicle, then you're going to fucking work!"

Prepared for this sort of comment coming from him, I replied "I never wanted to be on this vehicle but I didn't have any say in the matter."

"Welcome," SSgt Benson answered as he gathered his belongings.

Shortly after I had settled in, our Field Mess commodity set up the chow we had for our Easter Sunday lunch. We had two pieces of boneless chicken breast with gravy, dressing, green peas, peaches, chocolate and vanilla swirl cake, and a small cup of fruit punch. The servings were tiny, but enough to fill a shrunken stomach. It was definitely a nice change from MREs.

Afterwards, we prepared to tow two Tracks into the bay and swap out engines and parts in order to make one good Track. Just as we were about to hook up, M-03 broke down. So in the end, we were the ones being towed back to where we were. We decided to wait until the next day to find and repair the vehicles because it would soon be dark.

Cpl Daniels came over and talked for a while. It was fun listening to him tell stories for two reasons—first, he was attached to an assault unit so he had done and seen a lot; secondly, he looks and talks funny. His nickname is "Moon Beast" and he has something about him that makes him really likeable and easy going.

The funniest story he told was about entering the Presidential Palace. They had driven around forever looking for Saddam's palace but just couldn't find it. They had been driving past it without realizing it, and then noticed a fancy black car leaving through the gates.

They thought it might be Saddam or one of his sons, so they chased it with the Tracks only to get lost again. The top speed of a Track is only about 40 MPH, so it wasn't like they were going to catch this car. But they managed to follow it long enough to get lost in Baghdad once again looking for the Presidential Palace they were just at.

Finally, they ended up back at the palace with no way in as the gate was now closed. They called in the combat engineers to blow the gate. Apparently there was enough C-4 there to blow it off the face of the earth, because they wanted to make sure that when the gate did blow, they would be able to storm the palace and allow no one to escape. Just as the engineers were finishing up lining it with explosives, the gate opened. One of the troops had come up with a better idea; he had jumped over the wall and opened it from the inside.

"Come on in guys," the troop said as he waved them inside.

The next story he told was actually a horrible story, but there really wasn't anything else that could have been done.

They had just destroyed a pickup truck full of Iraqi combatants. All were killed except for two that managed to still be alive. One was so bad that he wasn't going to be living for long; the other was paralyzed from the waist down.

They tried to give them medical attention but there wasn't anything they could do for the one. The other, who was paralyzed, refused treatment repeatedly. They couldn't force him to accept so they just disarmed him and left him be.

He was trying to escape, even though he was paralyzed and they were not holding him prisoner. He kept dragging and pulling himself away with his hands and arms. Then he would try to place his legs underneath him and stand up, but he just kept falling.

A couple of times, he actually stood up all the way, but then fell right back down as he tried to walk. After watching him repeat this process for a few hours, someone finally made a joke.

"Five Dollars says he'll make two steps before we leave," bets a Marine.

"Ten Dollars says he'll fall trying to make his second step," countered another.

When he told us this, none of us knew how to react. This was very cruel, but at the same time, we could see how it could have happened after they offered him treatment but he refused. To make matters worse, those Marines couldn't leave their positions, and had to witness this struggle for life. And we all knew that to keep our sanity around here sometimes, we can't go too long without a laugh.

Then Cpl Daniels felt bad for even telling this story, so he told us another one about one of the Marines he was with ...

They were in a firefight and Sgt Belgarde was in the weapon station. They were taking so many casualties in this fight they actually had to turn back and start a casualty collection point.

After the battle was over, everyone was recuperating. About thirty minutes later, Sgt Belgarde was spreading some peanut butter on his crackers when he noticed he was getting blood on them. Blood was dripping down his arm, but he didn't know where the blood was coming from or if it was even his. His voice then calmly came over the net, saying, "Be advised, be sure to check each other over. I just found a bullet hole in my arm."

Sure enough, he had been shot twice from an AK-47. One bullet had ricocheted off his watch, and the other had lodged in his arm. He had such an adrenaline rush from the fight that he didn't even know he had been shot.

By now it was dark, and time for bed. But walking to my tent, I passed Nik and some other guys in conversation. I wasn't tired, so after setting up my bed went and joined them for a while. The conversation was about bikes, and we talked about the bikes we wanted to get when we got back home. I had been thinking about this, too. Most of them wanted a crotch rocket but wanted a big one like a Hayabusa.

I've always leaned more towards a happy medium like a 750cc. The 600cc bikes are quick off the start, but slow in the top end. The bigger bikes like the 1100cc or 1300cc are fast top end but slow off the line. The 750cc always seemed just right for me.

A Harley Davidson would be nice, too. You can never go wrong with a Harley, but they don't strike me as having the sport thrill that I would like in a bike. If I got a Harley it would probably be a Fat Boy or the new V-Rod.

The topic died out around 1700, and I went to sleep.

"Get up, Reveille, Get up!" Potato Head shouted the next morning (April 21). I cannot stand being woken up like this. I just want to lay there for spite. How irritating and what a bad way to begin the day!

I didn't like sleeping under the shelter on the concrete slab, either. It wasn't like I slept badly, but my hidden spot was much softer and I could see the stars. And I hate sleeping with the platoon because I have to wake up to shit like this.

At 0400, instead of going to check in, we had an accountability formation. This is stupid, too. We had no word to pass so why have a stupid formation?

Potato Head gave a ridiculous class on things he said we needed to know for the return home. He told us our wives and girlfriends were going to be used to having control while we'd been gone. So to prevent them from leaving us, we would have to retake authority and control gradually.

I couldn't believe this man was giving us pointers on relationships. Russ and Ryan were looking at me, dumbfounded. This guy has been divorced at least once, and no telling how many girlfriends had left him in the short time we'd known him. It's no wonder they leave!

After the formation, Ryan and I went back to begin working on M-03. She wasn't getting fuel but the fuel pumps were supposedly working. We changed out a few parts and still, she wasn't getting fuel. Then we tried changing out one of the fuel pumps even though they were said to be working. With the new pump, she ran great.

After that, Lope Dog told us we needed haircuts. I shaved my head again. When I was done, I washed my head with shampoo and water three times before I ever got all the dirt off. I also shaved my face, and then, fresh and clean, felt like a new man. It's amazing how the smallest things affect us out here.

Then Gy Foust came by looking for Lope Dog, but Lope Dog wasn't here. Gy Foust got mad because he had to walk so far to reach us, so he told us to move M-03 closer to the bay. We hadn't wanted to be closer because it made us more accessible to people we didn't want to be accessible to. We tried to buy ourselves a little time by telling him that M-03 had broken down and this was where we were towed. But then he asked why we weren't fixing it, and we had to tell him it was already fixed. Realizing what we had just tried, he got mad again; needless to say, we moved the Track closer to the bay.

Lope Dog came back with mail. I received a package of letters from an entire middle school Spanish class in Biscoe, NC. The students' cards were made from colored construction paper and were quite large.

I spent the remainder of the day just hanging out and visiting with different people. I ate a meatloaf with gravy and mashed potatoes MRE, which is now my new favorite. Another day down, but who knows how many to go.

On April 22, I woke up in the tent with Ryan and realized the day marked my five-month anniversary of meeting Abby. Unfortunately, I wasn't in bed holding her close in my arms. I must have said this out loud, because Ryan jokingly cuddled up next to me.

After we got up, we had a series of formations for various reasons. We turned in most of our ammo and kept just sixty rounds. We also turned in our MOPP gear which was great for two reasons—we didn't have to keep up with it anymore, and it freed up a lot of space in our packs.

Then I took a baby wipe bath. This is the next best thing to a shower, but there's no water involved. I wiped my entire body with baby wipes in the tent. It left me feeling sticky afterwards, but at least it took care of most of the smell.

I wrote Abby a Happy Anniversary note. Then I read a couple of chapters in my book. I am close to finishing Book 4 of 6, which will end Part II of <u>Lord of the Rings</u>.

A little sleepy, I climbed into the driver's station of M-03 for a mid-day nap. There's not a lot of room in a driver's station, but it makes for good upright sleeping.

When I woke, we had received more mail. Much to my surprise, as well as Ryan's and Russ's, Tammy had sent me a letter:

Eric,

Hey! It was nice to hear from you. You're right, I wasn't expecting it. I hope everything is going well for you!

Well, I've been so busy lately. I would have written earlier. We had Spring Break last week. We went to New Orleans (Mardi Gras) and Atlanta. We had a great time! Then this past weekend we went to Wilmington. It was weird because we stayed at Ryan, Carter, and Adam Browne's house.

So what's it like where you're at? I'd be scared to death if I were in your position!

I was very surprised about the card you sent. It was really sweet and almost made me cry. Then, in the letter you tell me all about your girlfriend. I wasn't sure what your motives were? I'm glad you and Abby are happy, but I was surprised you thought I needed to know. Not that I don't care about you ... just weird, I guess. I know we both have grown apart a lot, but I still care about you and always will. Me and Luke are very happy, but I'm sure it doesn't matter to you. It's just weird because you tell me about the promise ring and everything, but I hope everything works out between you guys.

Anyway, I hope you write me again to see how you're doing and what it is like out there. I love to hear from you and I'm sure you like getting letters, so as long as you keep writing me I'll keep writing you. Everyone in my family says Hello. Tell your parents Hello! Take care and be safe.
                    Miss you!
                    Love, Tammy

I also got a note from Dad dated March 17, one from Patrick, another from A-Rod (Aaron), and three from Abby. This is as good an Anniversary gift as any under these circumstances, although she did forget our last one. She was too busy chewing me out about Morgan.

                                    Monday night, 3-17-03
Dear Eric,

We were very disappointed to find out that we missed your phone call last night. We had stayed home the entire weekend thinking that you might call, but then Sunday night about 6:15, we took the Lexus to Asheboro to drop it off at R&D Automotive for Rudy to work on Monday. We got back about 8:00, and had a message on the answering machine saying "Hang up and try your call again later." I think your mother cried all night. I'm so sorry—I should have known better than to leave. We called Abby to see if she got a call and she had not. I think she was also out briefly, so maybe you missed her, too. I know you stood in line for hours and we let you down by not being home.

We just watched the President's speech at 8:00 PM giving Saddam 48 hours to leave Iraq or face war. No one thinks he's going to leave, so by Wednesday night, military action could start. We still hope for a small miracle and that someone would take him out somehow so no innocent people will be lost or hurt.

I can't imagine all the thoughts going through your mind as you wait for what's to come. I can only tell you that your mother and I think about you about every minute of every day. I know this is also true of about 150,000 other families, as well as a lot of other Americans. Never forget that your country supports you and owes you a debt that probably cannot be paid. Plus, be as proud of yourself and the Marines with you as we are of you.

It seems so strange to be writing you about things that will possibly be over by the time you get the letters. I hope and pray that this letter finds you safe and on the other side of the war with your thoughts more toward when you can leave for home. Please know that your whole family loves you. David called tonight to let us know that he is thinking of you. I don't think he even watches the news ... everybody handles it differently. I tell your mother that she watches too much news.

Please tell Ryan and Russ we are hoping and praying for them. Watch each other's backside and write when you can.

Love always,
Daddy

March 7, 2003

Eric,

Well, your letter got to me on March 7th (my birthday) and for that I am very thankful. I am glad things are okay here. I am not doing too well in school right now, but I am trying to turn it around.

That is great to hear about you and Abby. I am very happy for you two. All she does is talk about you. Me and Brandy are trying the dating thing right now. Abby probably told you. Today we are leaving for Panama City, FL. I am so excited. It is supposed to be warm and I hope it is. It is warming up here, too. I hope you're back for summer, wakeboarding, beach, lake—you know how we do it. As long as you come back, man. Just be careful! I am glad to hear that things are kinda calm over there.

You are definitely missed here. I'll tell you I don't envy you for the job but I envy you for the experience. Enjoy it, take it all in, but be the fuck careful. I am taking a history class on the Vietnam war. It is wild to spend 9 to 11 weeks just on that war. If it gets like that war, you get the hell outta there!

Anyways, bro, I gotta run; I'll write you again soon.

I love you, bro
Patrick

P-Rod,

What's goin' on, bro? I hope these letters are getting to you. So what's goin' on with you? Do you move around a lot or have you pretty much maintained the same location? Abby told me where you were, but I wasn't sure if you were still at the same place.

I miss ya, man—each weekend I always keep wonderin' where you're at. Things just aren't the same without ya here.

Things have changed a little in my life too. Me and Jess aren't living together anymore. We are still talking and hanging out, just not livin' together. I thought it would be best ... maybe we'll start missin' each other rather than arguing.

God, man, didn't shit seem so easy last year. Everyone was goin' out, smiling, havin' fun ... no worries. Now look at us. Actually, Abby misses your ass a lot. All of us try to keep her upbeat, but you're sure on her mind.

Kate had her going away party the other night—it was fun. I'm glad her and Russ got together. They both seemed happy in the pictures she showed me. Oh ... Richard's got him a new little girl that is 19. She's cool, but she's really got him—she is Jessica II. Ha ha! Na, but they are good together. Patrick is still the same good old Pat.

Oh yea, I started bartending at Phil's. It's a cool bar, but I'm still at Aqua as well, so I got that goin' for me.

Well bro, write back and tell me everything goin' on in your life. I miss ya. Take care of yourself.

Aaron

In her letter dated March 20, Abby had taken the time to sit down and answer many of the questions I asked in some of my earlier letters, especially the ones about her Caribbean plans and trip. Then in one note, I learned she had decided to get involved with some "Support the Troops" activities ...

Monday, March 31

Eric-

Hey Baby! How are you? I know you told me not to worry about you, but how can I not? Things here are okay.

I've decided that I am going to start going to the Support the Troops rallies. There are tons of them each week. I wish you could see how much everyone here supports what you are doing. Although we support you guys, we don't exactly support the war itself. So I think I am going to drive to D.C. to go to the biggest Support the Troops rally. I have to do something—I wish I could help you but I can't, so I'm going to do my best over here to help people.

So Friday night I got off work early @ 3 AM so I decided to do "late night." Well, we went to Austin's house—me, Jen, and Brandy and at first I was having fun but then Joe found out that his best friend had been killed at war. He received the call while I was standing next to him in the kitchen and my brain went crazy. I started crying—I couldn't stop it—just the thought of receiving a call like that drove me insane. That is my biggest fear—losing you! After that night, the reality of war has been completely different. Now I know I have to focus all my energy somewhere, if not I will really drive myself to insanity.

I never dreamed how hard this would be without you near.

Love you,
Abby

Her third letter was written April 2, and in it she sounded happy because she had made an audition tape for a new television show. She said that her roommate, Jen, was having her family visit that week— Jen's mom, dad, brother and sister—and they live in a two-bedroom apartment. She said Jen's family got up each morning around 6:00 AM, and that's usually about the time Abby gets to bed after working so late. I laughed to myself and thought it would probably be an interesting week.

Abby writes the best letters when she is happy, or at least not insecure. They are so perfect that I am just content with my life and I don't even feel the need to take action and respond to them, although I wish I could in person. She is so right for me that my biggest fear is losing her. I don't think it would ever be possible for me to meet anyone else who could bring me as much happiness as she does.

After the mail, Ryan and I made a homemade gym. We had four 5-gallon buckets with handles that we filled with sand. We used them to do all kinds of lifts, but mainly just curls. We also set up a place on the Track where we can do pull-ups, dips, and push-ups.

We were incredibly disappointed with our performances, or lack thereof. We're going to have to hit the gym hard when we get back.

After our poor workout attempt, Lope Dog had a good rumor for us. We could have a possible flight window of 15-31 May. Of course this is subject to change, but optimistically, it's only a month away.

I mean ... how long can a month be?

Chapter Twenty-Two

The next morning, April 23, things were becoming very habitual as we went to our 0330 accountability formation. Word was passed that we now had a shower from Supply, and it was mandatory that everyone shower to increase health standards. Lately lots of people had been getting sick.

After formation, Ryan, Russ, Lope Dog, and I began suspension work on M-03. We replaced ten road wheels and five torsion tubes and support arms. The process lasted all day with only one break for lunch. I received a letter from Dad while working, but didn't want to open it until later. When I got mail, I had to have private time to myself.

After we finished work, Ryan and I went to our gym and did a quick workout. Then we went to the maintenance tent to see if they were playing a movie on the Gunner's laptop, but there wasn't one.

So it was letter time, and I went back to my tent. Dad's letter was written on March 30 when they were staying with Kim. My niece, Julia, also wrote me a letter on the same paper from the pad Dad used.

Dear Eric,

We have been at Kim's now for two days and we're all doing fine. Yesterday, Josh, Julia and I hit golf balls across the street. Josh said he could hit the neighbor's barn (about 250 yards) and I said he couldn't. Well, I was right but he could hit it farther than I thought he could. Julia did pretty well, too. We also took the kids to Sir Pizza for supper.

Today Grandma Jones and I went to church. People there were all asking about you. It makes me feel good and proud that they all asked. We ate lunch at Jean's—the food was cooked by Lorraine for Stephen's 18th birthday, so all of her family was there. Larry and Kay, David, Kim and her family, and Tom were all there. Randy said the blessing as usual and said some very nice things about you. I have always liked Randy, but that made him mean even more to me. Everyone is trying to deal with the war and your being there in their own way. We heard today that some soldiers are getting mail and we hope that includes you.

Cynthia and I sleep in Kim's back sunroom on the fold-out sofa bed. Last night, it thundered and lightened all night and poured rain. I kept waking up wondering if you were hearing similar noises and seeing flashes caused by bombs. I wished I could just reach out and pull you into a safe place for a few days of relief.

Today, when we came out of church, it was snowing like crazy. It snowed hard for about an hour and turned much colder. Yesterday we could have worn shorts—this weather has been really strange.

Tuesday is Ryan's 21st birthday. Last night he went to Greensboro to spend the night with some friends and left the top down on his Jeep all night. Of course, the pouring rain drowned his upholstery and dash panel to the point that his Jeep had to be towed back to Asheboro. He is trying to dry it out now so he can go back to base tomorrow. Does any of this sound a little familiar? (Snow in Sam's driveway when you were to leave for California ...)

Tell Russ, Ryan, and Sgt Pit that we're thinking about them and hope they're doing well. It helps to know they are with you and you're all protecting each other.

We can't wait for the day you come home. I don't think I'll ever ask for anything else after that (I'm sure I will, but for now that's all we want).

Julia is also writing you a letter here beside me and it is enclosed. God be with you ... write when you can.

Love always,
Daddy

Dear Eric,
I hope you are safe.
I hope it doesn't last long.
We have been praying.
How are you doing?
I miss you.
I love you very much.
You are very special.
Love, Julia

_____

My parents built our Asheboro house in 1979—the year before I was born. It had been home for more than 20 years. In the months following my departure for boot camp and Dad losing his job at Black & Decker, they decided to look for a house they could purchase to renovate and resell. Dad had a B&D severance package and he and Mom had done some fairly extensive renovation on our Badin Lake cottage, so they felt this type of project might lead to a career option for Dad. Mom was working in High Point at the time, and found the perfect "worst house/best neighborhood" project.

For the next 10 months, they both traveled to High Point from Asheboro daily—Dad to his rehab project and Mom to her office. Then the company Mom worked for offered Dad a job. The job offer was both good and bad news—his severance was running out so he was going to be looking for other employment but the house renovations were not complete yet he was needed for the new job immediately. For a few weeks, both he and Mom worked during the day at Concept Plastics, then drove across town to work on the High Point house until 10:00 or 11:00 at night, and then made the 45-minute drive back to Asheboro.

Things seemed almost providential when some family friends whose house had burned down approached my parents about renting their Asheboro house, enabling Mom and Dad to move to High Point to live in that house while completing renovations.

But by the time all their rehab work was finished, they had become attached to the house. It was just the opposite of our Asheboro house, located in a great neighborhood on a 59' city lot. Mom was enjoying getting to decorate a "new" home after all those years, and Dad enjoyed the minimal yard maintenance. And there was an added financial benefit.

High Point, NC, is known as the "Furniture Capitol of the World." It is the home of the International Home Furnishings Show held twice a year. Furniture manufacturers and buyers from all over the world attend. Since most sales reps come for an extended stay, many prefer to rent private homes rather than stay in motels. The new house was located just moments from downtown, and was perfect for that purpose.

So they continued to rent out our Asheboro home and lived in High Point for about three years, moving out for a two-week period every spring and fall for the furniture market. But as time passed, Mom became increasingly anxious to move back to our Asheboro home. Dad, on the other hand, was quite comfortable in High Point and did not really want to have to make the drive back and forth from Asheboro to High Point every day.

The lease had already been signed for the spring, 2003, furniture market when our deployment date was set. Mom and Dad had planned to move in with either Kim or David for the two weeks their house was rented. The opening of the market almost coincided with the start of the war. Although moving out of her house and away from the phone was the last thing Mom had wanted to do, she apparently realized it was actually a very good thing …

Sunday, March 30

"… Jimmie and I are at Kim's house now. We got here Friday night. I can't tell you how much better it's been to be in the country again—to wake up and look out into the woods and sky instead of a street full of passing cars and the neighbor's windows. And the quiet—to actually sit on the swing and have the traffic in the distance and no dogs barking next door—and the privacy—well, you know exactly what I mean. It's amazing how much the setting of this house reminds me of ours. It's no wonder Kim fell in love with it. I close my eyes and see you riding across the field on your bike—with no helmet, of course."

Ryan came back to the tent after a little while with a *Truckin'* magazine for us to read. There was a Chevy Silverado that Oakley had bought and customized as a show truck for the motocross enthusiast. We compared it to the Hummer H2, and decided that the H2 was better.

The next morning, April 24, I was informed at the 0330 formation I was to be the Corporal of the Guard for the day along with Moya. We were also going to be able to turn in our Atropine injectors. It was good being able to get rid of these, because they were all accounted for and we didn't want to lose any.

Atropine injectors are to be used in case we're hit with a chemical nerve agent. If we're hit, we are to take out an injector, press the button, and then jam it against our thigh as hard as we can. A pressurized needle would shoot into the meaty portion of our thigh and release an antidote for the nerve agent. The only downside to this injection is that if we weren't actually hit with a nerve agent, the Atropine itself would kill us.

At 1000, I took a shower with three gallons of cold water. The shower is a hanging bag of water with a spout at the bottom. We put ours inside the engine compartment of one of the Tracks. When the plenum or air induction is lifted up, there's a lot of space and privacy.

The only bad thing is that the floor is sloped toward the engine, so we have to stand barefoot on the rails for stability. I slipped off the rail a couple of times, and it's really frustrating to stump a toe on the engine.

Even though I wasn't really clean, I felt great afterwards. As dirty as I was by going thirty-eight days without a shower, I couldn't get clean with only three gallons of water. But I was a lot cleaner than I had been before.

At 1100, we ate hot chow that was so bad it messed up my stomach. Good thing I have post all night; I can shit in the middle of the night without anyone resting their back on me.

I got mail again today and a lot of it. It's another good thing I'm on guard, because I'll have nothing but time to sit in the maintenance tent reading and writing letters.

I received a card from Sam and Sansia Coble, who are my relatives somehow. Sam is my Grandma's sister's son and Sansia is his wife so I guess that makes them my parents' cousins, which would make them my 2nd cousins. I think? Aside from that, Sam is a lot of fun to be around because he's always got a joke to tell. Sansia is a counselor at South Asheboro Middle School so she always had her eye on me.

I also got a nice letter from Jim Davis who's in support of the war and a letter from Morgan along with two pictures of her.

3-27-03

Eric,

Hey Babe! What's up? I got your letter today—I think you will regret saying "write as often as you can" because you'll get like 2 a day. J/K! So how are things? Good, I hope. Thanks for the picture. I showed my friend Christina who has been asking me to get her an address for one of your friends because she wants to write to a Marine, because you guys look so cute.

I'm going to send you sweets but I don't know if I should get them from my bakery because I think even with the bread in there they will get hard. So I'll send something safe. If you haven't gotten it yet, you'll probably get it in the next few days.

Anyway, I thought you would probably laugh out loud when I told you about Jeff. Thanks for your advice on that situation. A lot of it rang true, but I'm not afraid to be alone or afraid of losing a physical relationship. It's just that Jeff is a wonderful guy and I don't have any negativity towards him. The hardest part is that he has become one of my best friends. I think that is what he and I are best at ... being friends. It's so hard to break his heart because I care about him so much. But I know he won't be my friend when our relationship ends. He'll be too hurt and have to move on.

Here I am just rambling on but I'm sure you understand what I mean. I hate being the mean one, and it's hard to say good-bye to someone I care about and know how much I am hurting him at the same time. But I'm not going to find the right guy while I'm dating the wrong guy and I shouldn't settle for less than what I really look for in someone.

So for now I am going to take time to be single and live it up with my girls.

Anyway, I have a funny story. My little sister who is 11 years old (McKenzie) has a boyfriend at her new school—cute, huh? Well, her boyfriend and all of his friends TP'd our house this past weekend. OK, that's not really important, but we cleaned it up and everything and on Monday morning the doorbell rang and my Mom and I went to answer it and outside was a fire truck and six firefighters in biochem suits and gas masks around their necks. And they asked us if we got toilet papered this weekend; we said yes, and they proceeded to tell us that they had gotten several calls from people in our neighborhood because some flew down a couple of streets and was wet and got squashed in the road. It apparently was not recognizable as toilet paper and they thought it was some kind of biohazard resulting from a terrorist attack. Can you believe that? How insane do people get? I'm positive that if there was a terrorist attack it would happen in this little beach community of S. Redondo!!! The firefighter didn't even crack a smile. He was completely serious.

So that's my story—my neighborhood is crazy!

Well, I still have no idea where you are. Iraq? Kuwait? Wherever you are, I hope you are safe. I'm thinking about you! Write when you can.

Love always,
Morgan

Kim mailed me a large envelope of cards made in Julia's Kindergarten class. There are also pictures of them outside the school by the flagpole where they sang "God Bless America" and "Proud to be an American." There's also a newspaper clipping from The Courier Tribune in Asheboro where Kim had my name and address published in the paper. Mom sent me a letter and sounded better than she had but it was obvious that she's stressed.

And I got six letters from Abby; they were dated March 21, March 22, March 23, March 24, March 25, and April 4.

I read her first letter, and began the second ...

Saturday, March 22, 2003

Eric—
Hey honey. So the last letter I wrote I said that I was mad at you—but I'm not really—I'm just worried who Morgan is and what she means to you. I'm sorry I was mean in my last letter. I do love you.
Yesterday, which was the second day of the war, I went to visit your mom. We sat in the sun for like 5 hours and talked about you. We talked about your trip home from California, when you were in middle school, and everything else in between. It's comforting because she misses you as much as I do. It was the first time all week that neither me or your mom watched TV all day long. When your dad came home he tried to join our conversation, but he acted like he felt uncomfortable so he went to the bank so your mom wouldn't have to go. And when he came home again we decided to go to dinner. We went to your favorite restaurant, Liberty. It was good and I liked the restaurant's interior. Your dad said you would be mad that you didn't get to go with us, but I think you would be glad that we got to go together. I do wish you could have been there.
While I was talking to your mom, she made me feel like she thinks we are rushing into things. I don't feel like we are, but on the other hand, I could not imagine anyone else more perfect than you! You are perfect for me, and I do know we haven't been dating that long, but I feel like I honestly know you and you know me—so why does it even matter? I feel like I could spend the rest of my life with you. It's like, if

there was one person made especially for me—that would be you, Eric. I don't know about love or marriage or what it takes to make a relationship last a lifetime, but I do know that we are such similar people how could it not last forever.

Enough about all of that! Just hurry home so we can begin the rest of our lives together.

Your mom called me today to thank me for coming up yesterday. Last night, when we went to dinner, that was the first time your mom had been out of the house since Sunday night. She worries me, but if I didn't have to go to work or school, I would be the same way. I don't like leaving the TV because while I'm gone they could mention Camp Matilda or anything that I wouldn't know about.

Well I miss you and I'm depressed. I really need you right now. It would all go away if your arms were around me. Dreaming of you …

> *Love U,*
> *Abigail*

I couldn't believe it. Abby apologized to me on the very next day after she blessed me out. Our fucked up mail system caused me to lose faith in the girl I loved. On top of that, I had sent a hateful letter to her because I was certain she hadn't apologized. She was probably just now receiving that letter, or would be shortly. She won't ever talk to me again. If I was at home or had access to a phone, all of this could have been avoided. Oh, well.

Monday, March 24, 2003

… My mom and sister are visiting now and it's going well. My mom and dad are trying their best to be supportive of me. Every time I find out about where you are or what you are doing, my mom writes it down. She's trying to show me how much she loves me and you together. Mom has a hard time being supportive of me but she's doing her best. But you already know all of that!

… Aaron called me at 4:30 AM in a panic—he is so worried about you. I had just fell asleep at 3:30 AM so I was delirious when I was talking to him, but he was saying something about Iraqis blowing up a bridge and 40 men were killed …

… Kate's friend Shawna came into Mythos Saturday night and let me read a letter Kate had written her. Kate told her to give me her address so I could write her. I think I'm gonna ask Kate about that night to see if she did anything with that guy. I don't know

whether she will tell me the truth or not. What do you think? Should I ask her? Or no? ...

Friday, April 4

Eric,

Well hello beautiful! How are you doing? Are you okay? I'm worried about you, as always. I'm sorry I have been so selfish lately, it's just when I had a problem you were always here to give me advice, but this time I am all on my own ...

... Yesterday President Bush came down to Camp Lejeune. Jen and I left Wednesday night after work and drove out. We got there at 9 AM. We were so excited we got to "support our troops," but then we couldn't even get to see the President. They handed out tickets between 7 — 9 AM at the VFW, but we didn't even know you had to have tickets. Well, next time we'll know better. We bought a paper from Jacksonville that told us everything. It was a fun trip and it brought back a lot of memories so it was therapeutic. I need to remember you. I think that's what scares me; I don't remember your scent. I think a lot about when you kiss me and what it feels like to breathe in and have your air in my lungs. I miss that. I miss everything about you. I need you and I look forward to the day you come home.

So Jen and I just got home—she got her nipple pierced. I want mine done, but I want to know what you think about it. Do you care? Can I do it? Should I do it? I really want to. Will you be mad if I pierce my left nipple? I just don't want you to think I made big decisions without your input. It's hard, but I don't want to make big decisions in my life without you. It sucks.

I finally got your Valentine's Day card. I cannot thank you enough for everything you said. Thank you for my poem. It was beautiful. Words cannot express what it meant. Thank you for loving me even when I'm not worth being loved and even when I'm selfish.

I love you always,
Abby

I had some free time, so I ventured out in the dark to take a shit. I was sitting there all alone when I heard the brush rustling and crunching as if steps were being taken. I assumed someone was coming to join me or maybe they were just going to take a piss. I looked around in the direction of the noise but couldn't see anyone, so I began to worry.

I had my rifle with me but no rounds, so I pulled my knife out. I flicked on the small lantern light that Abby had given me but it wasn't bright enough to see more than five feet or so.

Then, about 15 ft. away, I saw the light reflecting off a pair of eyes—reflecting with the kind of greenish glow animal eyes give off in the dark. Guessing it was a dog and hoping to spook it, I made a deep growling sound. It disappeared.

This was funny to me on two levels. First, I had just growled at something to leave me alone. Secondly, I might've had to fight something with my pants down. It seemed bad that I needed to carry a loaded gun with me to take a crap in the dark.

It was now past midnight on the 25th of April. At 0100, I finally got to take a nap before morning formation. It was always all I could do to wake up for formation, and I knew that without a nap, I would be worthless all day.

SSgt Ferguson came over with a bad gasket on a PTO. Ryan and I worked on fixing it with the help of a few floor mechanics.

Then I skated away into the driver's station and took another quick nap. I woke up hot because the engine was running and the driver's station is beside the engine. They were just finishing up on the PTO, so I helped with the final touches.

I found a *Truckin'* magazine when I got back that had a special on the Hummer H2. I want one so bad I can already see it. Black with 37" x 14" all-terrains mounted on 18" chrome wheels.

Maybe I could buy a used one when I get out? I could work two jobs and save my money from now until then. O.K., enough daydreaming.

Then I got mail from Marie and Ernest Pugh, who knew me from high school. Evelyn Yow, who knew my family. And letters from Tammy and Dad. Dad's letter was written on April 2, when news reports about the war had become a little more positive, or at least a little less terrifying for them. Tammy's letter seemed a little warmer than her first …

March 31, 2003

Eric,

Hello! So how are things with you? I hope O.K. I'm so worried about you. I'm not sure you got my last letter, because I wrote "AAN" instead of "AABN," but I hope it got there. It sucks that it takes so long to receive letters.

My Mom sent me a picture of you that was in the newspaper. You looked so cute! I called your parents and talked to your mom. She was so upset and worried about you. She said she was glad I called. It was so good to talk to her—she was so sweet. I just love and miss your parents/family so much. I know you and I aren't really that close, but I still love you so much. It hurts me so bad to know you're out there. Even though we don't talk much, I think about you all the time and I am so scared for you. I just want you to hurry and get back, so that I know you are O.K. I really wish I could have seen you before you left, but I understand your circumstances. You have just been a huge part of my life and such a special person to me. Every time I watch TV, I get tears in my eyes because I am so worried about you, and think to myself I can't believe this is really happening.

Well, I am praying for you along with many others. Please, please, please write me back when you have a chance. I know you are getting tons of letters and I'm pretty sure I'm not your first priority, but I really want to hear from you. Keep me updated on where you are and what's happening. I just want you to know I'm thinking and praying for you! I love you so much and hope to hear from you soon. If you don't have time to write me, have your mom call me or something!

Be safe! I miss you!

Love you,
Tammy

I understood from Dad's letter why Mom had sounded better in her last letter, and he told me about a visit they'd had from a neighbor.

"... We think the people in our Asheboro house are leaving the end of June and we are probably moving back home when they do. I think we might try to sell the High Point house and just pay off the other two. Cynthia is looking forward to getting back home and in some ways so am I. We might even be back by the time you get home.

... Something else happened last night. We had not been home long—I was in the den and your Mom was in the bedroom when someone knocked at our front door. It was raining, and as you know, no one knocks on our "front" door. Well, we both panicked and ran to the door. It was not military (Thank God!), but instead was one of our neighbors and their little girl. The mother and daughter (about Julia's age) were standing in the rain bringing us homemade cookies. They said they had found out that our son was in Iraq and just wanted us to know they were thinking about us and wanted to show their support for you. We both were speechless and deeply moved. This is exactly the kind of support that we see everywhere."

He also mentioned the rescue of PFC Jessica Lynch. By putting two and two together, I realized he was talking about the female POW and terrible rumors we had heard as we traveled through the An Nasiriyah area when the war first began. It seemed strange that I was learning details about the war from him and others writing from home when those of us who had been in the middle of it had been so starved for information.

And he sent me a quote from what he said was an e-mail he had received from Camp Lejeune written by our Commanding Officer dated 3-27:

"You can be proud of those superb men. Watching them do their jobs makes me enormously proud to be their leader. These are some of America's finest citizens, full of spirit and support for each other. My job is easy. It is kind of like being the Duke University Basketball Coach—just let them play and they'll make us all look great. 'Semper Fidelis!'"

Chapter Twenty-Three

April 26. The weather is beginning to warm up. Let me rephrase that. It's getting hot. Sleep last night was terrible. It was so hot that I kept sweating just laying there on top of my sleeping bag. Somehow, I finally drifted off.

After our 0330 formation, I went back to my tent and began responding to all the "fan mail" I had begun to receive from supporters. It was too hot in the tent, so I moved out to M-03.

AAV mechanics are generally attached to an AAV R-7 A1, which is easily recognizable because it has the boom on the top and there are only a dozen or so in existence. So if an AAV mechanic sees an R-7, the chances are good they know someone attached to that vehicle because there are so few of us.

And that explains why Alex Munoz from Phoenix and Justin Gantt from Columbia, SC, came by to visit. I went to AAV School with them out in Camp Del Mar, CA, and had partied with them quite a few times in Tijuana, Mexico. I also flew home with Alex to Phoenix and met his family over a long weekend. Alex went to Okinawa, Japan, after AAV School whereas I had gone to Camp Lejeune.

Seeing them again and catching up on old times was great. Alex had a new nickname, "Mine Plow." Thinking it probably was related to some kind of funny story, I had to hear how this name came about. But my assumption was wrong; there was no humor involved.

Alex was a driver of one of the AAVs through the war, and at one point had gotten lost. The HMMWV in front of him crossed the bridge over the Euphrates River fast and hadn't waited on him.

They had just made it across the bridge and as they were trying to navigate their way through the darkness, "Boom!" The explosion was them; they had just taken out two land mines. Luckily, the Track was still mobile and no one was hurt so they pressed on. Alex knew that the Track was damaged badly, though, because his steering kept getting looser and looser. But they were still lost and trying to catch up to the convoy.

About that time, the whole portside track fell off and sent them into a spin. Just after this, the convoy they were trying to catch came pouring past them in the opposite direction. They had missed a turn a ways back, so now Alex, his crew, and troops sat stranded all by their lonesome in the dark.

"Be advised: Stay alert!   All movement north is enemy movement!" came across the net. They looked north and could see headlights coming their way. They opened up fire from the weapons station and began destroying the enemy. But they were outmatched and the enemy kept advancing

The troops in the back of the AAV were all cooks, and not experienced in this type of situation. They dropped the ramp and charged out the back. They held their own as best they could. Just in the nick of time, a contact team pulled up and began laying down heavy fire. The firefight was soon over and Alex's crew assembled back at their vehicle where they slept for the rest of the night.

The next morning they were woken by swarms of flies buzzing about them. When they got up to assess the damage to the Track, they realized why the flies were there. They were feasting on dead Iraqis that had been ripped apart all around them. The stench was horrendous as the gases released from the shredded bodies. It was a haunting thought.

After they left, I couldn't get back in a writing mood, so I thought I'd take a short nap. I actually slept four hours and by the time I woke up, it was bedtime. I couldn't go back to sleep, so I went to the maintenance tent to see if anything was going on.

Kammerererer was in there, along with SSgt Little and SSgt Thompson. They challenged us to a game of Spades. We played a good game for a while, but they ended up winning.

When I got back to the tent, Ryan was still awake. He couldn't sleep because it was so hot, so we stayed up and talked a while about song lyrics we knew. We were so over this place and ready to go back home. Maybe trying to remember song lyrics would put our minds in better places?

We couldn't remember many of the lyrics to the songs we thought of without half singing the melodies as well. Then we thought about the song "It's Your Love," by Tim McGraw. Ryan insisted that I sing it, since I told him Abby and I had adopted that song as "ours." It just didn't seem to have the same meaning when I sang it to Ryan! I can't even begin to carry a tune, but who cared? The laughter helped, and we were finally able to go to sleep.

April 27. This morning at formation we found that we were going to have to give M-03 back to 3d Tracks. We cleaned out all of our belongings and moved them to our tent. It was sad to see her go, although it felt good to think that by this action we must be coming up soon on the list to leave.

At 0530, Ryan, Russ and I went to eat hot breakfast across the street. It wasn't very good but it was edible. We received one waffle about three inches square, some kind of egg and sausage mix, some red stuff, some white stuff, and some dry cake that I couldn't force down.

On the way back to our bay, I found a plastic bat and ball. So we went into the bay and played for a few hours until the ball got stuck on the roof ... game over.

The bay we were in was about the size of four small aircraft hangers all butted together. Actually, by the way the roof lines were, I thought they had done just that. There were a few wall panels, but most were missing. The roof had hundreds of tiny holes all over it, and larger sections missing. It made me wonder if this used to be a military outpost that had been destroyed in the war.

It was time to write a letter.

THE COURIER-TRIBUNE
Wednesday, May 21, 2003

Sister wants to share letter from Marine serving in Iraq with readers

Editor's note: The following is a letter from Cpl. Eric J. Cox, who is serving with the U.S. Marines 2nd Amphibious Assault Unit in Iraq. His sister, Kim Farmer of Asheboro, wanted to share it with readers. The letter was written to the kindergarten class at Farmer Elementary where Kim's daughter is a student. The class had sent letters, photographs and a care package to Eric and his fellow Marines. Kim said Eric also enjoyed the letters he received as a result of The Courier-Tribune's publishing his address in a special "News From the Warfront" segment.

April 27, 2003

Dear Mrs. Beane and Class:

I want to thank all of you for the letters and pictures that you sent to me. I passed them around to all of my fellow Marines, including my good friends, Ryan and Russ.

Everyone was excited to read them and astonished to know that we have that much support from you and everyone back at that wonderful place we call "home." It brought tears to my eyes when I understood that the whole class sang two songs, one of which was a big part of my history.

Upon completion of Marine Corps boot camp, a ceremony is held to present us with the symbolic Eagle, Globe and Anchor signifying that we had earned the title, "Marine." During this very emotional ceremony, the same song you sang, "I'm Proud to be an American," plays throughout the event.

About the war — no one wants war, but we as Americans have to be willing to fight when our freedom is challenged. We have indeed accomplished our mission and overthrown Saddam and his regime.

It did not come cheap, however; we paid the dear price of American lives. As my Commanding General told us before the war, the war and reasons for it can be summed up in two words — "never again." Never again will our country stand for any acts of terrorism or anything else that would endanger the lives of the American people.

Well, I must be going now but maybe one day soon, I can come home and meet all of you in person. Julia, you know I'll be home to see you regardless. Please don't worry about any of us out here. This is our job and we are trained well for it. Just keep in mind when you see us in action that there is "no better friend, no worse enemy than a U.S. Marine."

Also, since you all are into music, a song and desire comes to mind often out here. It is by Ozzy Osbourne titled, "Mamma I'm Coming Home."

I am sure you have learned from my sister, Kim, that our mother (Cynthia Cox of High Point, formerly of Asheboro) is terrified for me and will not rest easy until I return home. I wish you all the best of luck with everything you may encounter throughout your lives.

I got mail later on and just as I did, a huge sandstorm came in. I got a statement from Navy Federal Credit Union saying they are still withdrawing $200 a month for payment on a credit card I no longer have.

Colleen, a girl I used to date while she was in Charlotte and briefly when she moved to Austin, sent me a letter. She's in Graduate School at the University of Texas at Austin now, studying to be a neurosurgeon.

I came across a *Cycle Trader* magazine and went "shopping" for a little while. I made up my mind that I'm going to a Yamaha dealership when I get home with the $7,000 I will make while here and drive away on a new YZF R-6. They have a new silver and black color scheme that I'm sold on.

O.K., enough dreaming. It's time to go to sleep in this awful sandstorm. The good thing is that it isn't quite as hot with all the wind and sand in the air.

April 28. With the sandstorm still going strong, I started writing letters because I was way behind. I was determined to respond to all the people who had taken the time to write me. But after hours of writing, my wrist had all it could take and I found another magazine. This time, it's *Boat Trader,* and I searched for my new boat. It would be really cool to find a nice powerboat under $13,000 to play with. Especially if I moved to Wilmington! But I could commute on a bike.

As I was reading, Summers, Gene, and Khadiev came to visit. I went to AAV School with all of them as well, and probably spent one too many nights with them in Tijuana. There didn't seem to be a whole lot to talk about, though, so they didn't stay a long time. But it was great they came by.

After they left, it was naptime again. The sandstorm had finally subsided when I woke up. I finished up a few more letters, but then it grew dark and I had to stop writing.

April 29. This morning was very eventful. We had our routine accountability formation, but then learned we were going to a formation with all of 2d and 3d Tracks. Colonel Dunford, Commander of RCT-5, wanted to congratulate all of us on a job well done.

After the formation, I went back to my tent to get the letters I had to mail. I mailed all ten of them, and then began reading some in my book. It was putting me to sleep so again, time for a nap.

I take a nap every chance I get out here. At the first tiny sign of weariness, I take a nap because sleeping helps pass the time. Plus, I get to dream of Home.

When I awoke, I tried to catch up in my journal but after writing just one sentence, we got mail. I received fifteen letters. Two were from Abby, one from Dad, one from my brother David, one from my friend Brandon McKenzie, two from Grandma and PaPaw Cox, one from my Aunt Lorraine (Newsome), one from my cousin Laura (Newsome), one from my "Punk Rock Mama" Shelly LaCoss, another from Sam and Sansia, one from my 8th grade teacher Mrs. Martin, a letter and package from Kim with flea collars, and two from supporters Lauren Kilby and Jessica Evans.

Before I started reading their letters, I put on one of the collars Kim sent me. Then it struck me … I couldn't believe the new low I had just hit. I was wearing a flea collar around my neck. It gave a whole new meaning to the phrase, "Devil Dog."

It was great to hear from Kim …

Eric,

It is Sunday, March 23. I got your letter just yesterday. I had wanted so badly to hear from you, but the picture you painted saddened me. The news is filled with the terrible events of the war. I'll have to say that I have never felt this helpless in my life. I am so scared for you and want so to be able to do something. In this world we have gotten so used to picking up our cell phones and communicating anytime, anywhere with the ones we love. Now when it is the most important time to just hear your voice, it is silent. We are bombarded with news and events and yet the only thing I want desperately to hear is that you are OK. Please be OK! I keep questioning how this could have happened. How, at this time, in this war, did you end up in the middle? You always did like to be right in the action and the center of attention, but this is ridiculous. My heart is with you always as I find it hard to concentrate on anything else.

God, just get through it, be tough, be brave and be home soon. The life you are living now will soon be a memory and you can come home to open arms, loving family, green grass, rolling mountains and flowers blooming (not to mention big trucks and beer weekends) and the life you are supposed to live.

I'm sure you know, but I love you. I wish I was better at putting it into words, but when you were born and since I was

*older, I always wanted to take care of you. I still do. (I hope
you have gotten the candy and goodies that I'm sending—there
will be more to come.) But in this war, I know that you are not
that little kid brother. You are a strong US Marine. I am filled
with pride, and scared to death at the same time.*

    Please stay safe.

I love you, Kim

I finished reading all my letters and began writing replies. Then after a while I began to look at some newspaper clippings Mom had sent me.

I studied the stock quotes and tried to see some sort of pattern. But I only had the quotes from one day. So I tried to see a pattern in the long term. I'm at a point in my life where I need to begin planning for my future, but where should I put my money?

I read a magazine article that helped me clear this up. According to this article, the first thing I should do is purchase a home, since this investment has proven to be the greatest vehicle to establish long term wealth. There are tax benefits that allow me to pay more per month and actually save money in comparison to renting. And I would be able to build equity in my home. Equity is the difference between the market value of a home and the payoff on the loan. So the longer I own, the more I pay down on the mortgage. The more the home appreciates in value will also substantially increase its equity. On top of that, I have to live somewhere so why throw my money away to a landlord?

Once I become a homeowner, I should begin to diversify my investments by taking five to ten percent of my income and putting it in vehicles such as mutual funds or investment property. The way I interpreted this is to first put money away in mutual funds that earn a higher rate of interest than a savings account. Once there's enough money built up there, I would have the down payment for a rental property. Then the cycle would begin all over again.

This all made perfect sense to me. Why weren't more people doing this? Maybe it just hadn't crossed their minds. Or maybe like me, they wouldn't know unless they were reading some boring investment article. I went to bed excited about going home and starting the search for my first house.

At this morning's formation, I learned that I was the COG for the day with Nova. We rotated four hours on and four hours off so I was either sleeping or writing.

At midnight, May 1st, I typed a letter on the Gunner's laptop to Mom for Mother's Day. I wasn't supposed to touch his computer, so writing and printing a letter on it was enough mischief to brighten my spirits.

The card I sent her with the letter was a card that she had actually sent me earlier. It had a little stuffed bear on the front, holding an American flag and at the bottom said, "Red, White and Blue without you ..." Instead of writing on the card itself, she had written on Post-It notes and stuck them to the card so if I wanted to, I could send the card back out to someone as my own.

At 0400, Ryan relieved me as COG and told me there was to be another stupid formation at 0445. I climbed in my tent and zipped up so no one could see me. I escaped the formation and fell asleep for another four hours. Then I mailed the letters I had written yesterday. There were seventeen of them, and each was more than a page.

A little later I got a package from Morgan. She had sent Kool-Aid, beef jerky, Pringles, chocolate chip cookies, Abba-Zabba candy, and a camera. I'd never had this Abba-Zabba stuff, but it was pretty good. It's like taffy with peanut butter in the center. She also sent me a short letter and a picture postcard of the city of Redondo, CA, where she lives. She jokingly asked me when I was going to call her to tell her I would be moving there.

> Hey Eric!
>
> I hope this all gets to you! Tell Russ and Ryan I say Hi and don't forget to share with them. OK, you don't have to if you don't want to—I know I can only fit so much and still have it get to you. Alright, well enjoy!!!
>
> I'm going to go lay out and get a tan in the sunny hot rays of California! Are you jealous?!! J/K! OK, write soon! Miss you ...
>
> Love always,
Morgan

After eating and distributing the cookies, I reached for my book. We would be leaving soon so I wanted to be closer to the end. I read 50 pages and finished the second part.

Chapter Twenty-Four

One by one, agonizingly slowly, the days passed. They were getting longer and hotter and it seemed as if every day was a letdown. With the fall of the Saddam Hussein regime, our mission was complete. Our Tracks had been given to other units; there was no work to be done. All we could do was sit and wait for a ride home.

I remember back when I was a kid we would start decorating the house for Christmas right after Thanksgiving. The decorations were a constant reminder that Santa Claus would soon be coming with presents.

The problem was that I had to wait almost an entire month for him to come. Presents began to appear under the tree with my name on them, but I couldn't stand not knowing what was inside and having to wait so long to find out.

I would go to school during the day and play afterward. Then I had supper with the family and fit my homework assignments in somewhere. (Well, at least that's what my parents thought I was doing.) The routine helped keep my mind off Christmas and the gifts for the most part.

But then about a week before Christmas, school would let out until after the New Year. This always made for the longest week of my life. With nothing to do during the day other than play, I began a daily countdown, starting with "Only five more days before Christmas." Time almost seemed to be standing still during those last days, but when Christmas finally arrived, it was the best day of my life, every year.

When we get to leave this dump and go home it will feel like Christmas morning did so many years ago. The twisted problem is that we don't know when that day is going to be. So every day we wake up wondering if today is going to be Christmas day.

Time stands still in the worst place on earth. Morale is sinking and frustration is growing.

The best thing we can do is keep our thoughts as far from home as possible but that's difficult when it's the number one thing on everyone's mind. Sometimes reading can block it out temporarily.

There are a lot of magazines around that can help, but people hoard them and we have to really know someone in order for them to even think about letting us borrow one. And even if they're willing to consider loaning it for a couple of hours, they aren't willing to do it without a profitable trade for something of material value.

Talking to friends would obviously seem to be a good way to help pass the time. But the only thing we want to talk about is what we're going to do when we get back. So talking to friends is actually counterproductive because it ends up making us long for home even more.

Like sleeping, writing letters passes the time quickly. But it's one of the worst things we can do when trying not to think about home. And as I try to write about things I've seen and been through, what's on my mind is the possibility that I'll be home before the letter gets there. It takes three weeks for one of my letters to reach its recipient. If we were to leave anytime over the next two to three weeks, then there was a good chance I would beat my own letter home. So what's the point?

With so little to do and so much time to do it, we try to pace ourselves to allow for as little down time possible. We move much more slowly than usual.

I take my time dressing and getting to the morning formation. After formation, I don't immediately run off to hide like I had done before; I stay around to see if anyone has anything interesting to say or do. Then I walk slowly back to my tent to find my hygiene gear so I can shave and brush my teeth. I see if any of my buddies would like to go with me to do the same thing, hoping I'll have to wait for them to get ready.

Then I take my time walking over to the water bowl. I get there hoping I'll have to wait to fill my canteen cup with water in order to conserve our supply while I shave. And with all the time I've spent brushing, my teeth have never been so clean.

By the time I'm done with my hygiene routine, I'm hoping there's a long line for breakfast. The longer I have to wait in line, the faster the day goes. But I don't want to be too late, or I won't get any food. I have to time it just right. And even when I'm getting food, I have to be that guy that everybody likes so the Marines serving me won't give me a disappointing portion.

After I get my food, I turn around to wait on my buddies that may or may not be behind me. I take my time finding a spot on the ground to sit. I'm careful to observe the ant population, since I know I'm going to be there for a while.

When I finally get situated, I begin eating my food with the smallest nibbles I've ever taken. I take time in between bites to look around and see what other people are doing and if I can tell they are trying to do the same thing I'm doing.

When I finish, I continue to sit there and let my food digest itself. After I get up, I gather all my trash and throw it away. I'm not done yet, though. I patrol the area to look for other trash that someone may have overlooked after they finished their meal.

As I walk back to the bay and my tent, I look around for the longest possible route between the two points. I never know what I might find by taking a new path. Maybe I'll run into someone I haven't seen in a while, although I really don't feel much like talking.

By now the heat is so bad I have to get out of the cammies or coveralls I'm in. The problem is that I'm not allowed to be in PT gear unless I am actually doing physical training. So the trick is to lay in my tent in my PT gear with both doors open in hopes of getting some sort of breeze. The downside is the flies.

Although I didn't think it possible, the flies are multiplying exponentially since we've now been here for quite some time. They are attracted to the water we use for shaving and brushing our teeth, and to the dead skin floating around our tents and living area. Although we have daily clean-ups in an attempt to control their population, the longer we stay, the more they come.

There is usually a game of Spades or some other card game going on in the maintenance tent. But that's where the Staff NCOs generally hang out or occasionally pass through. They are just as bored as we are, but get kicks out of picking on the lower ranks. If I don't want to get told to do some sort of useless job, I want to steer clear of this area.

Our toilet facilities are now different. Instead of the "two-man" boxes we used to have, we now have makeshift stalls built from 2 x 4's and plywood. There are 4 of them side by side and connected to each other. It looks like a row of wooden Port-a-Johns. The target containers are underneath the seats and accessible from the open backs of the contraption.

These metal wash buckets have to be emptied once or more daily. The lowest duty anyone can be given is "Shitter-burning" detail. The waste contents in the buckets have to be burned by pouring diesel fuel into the bucket and igniting it. But it can't be left to burn by itself because the fire will go out and there will still be waste left. So while it's burning, the contents have to be stirred with a large stick or rod as if brewing a tasty stew.

Fortunately, I was never the one "burning the shitters," but I did have to oversee a few troops that were. This is the very reason I steer clear of the bored Staff NCOs. They walk around in search of someone doing something they shouldn't be doing, which in most cases was actually nothing at all.

For all these reasons, I was glad I had not read more in my book than I had. I was at pg. 730 out of 1,000; I still had quite a bit of reading to do.

We began receiving care packages that had been mailed as early as the first week of February. I guess the handling of the packages took longer than the letters. So now we had an abundance of goods that no one could even use.

We really needed the magazines in the care packages to help us pass the time, but had extra food that no one could eat. We weren't doing anything so we weren't hungry. Had these packages reached us about a month earlier, there would have been a great demand for the food and little demand for the magazines, since we didn't have food during the war and lost the magazines we did have. But it seemed that most of the magazines had been delivered during the war and all the food after.

It was sad to have to throw food away. If there was homemade stuff there was a good chance it was old by the time we got it. We'd open it up and try a bite; if it was good, then everyone would share it. Homemade is the best when it's good. If it was bad, then it was the first to go.

The next thing to go was food that might have been shipped in the same package as soap. Even if packaged food and packaged soap was double and triple wrapped in separate freezer bags and placed in opposite ends of a box, the food tasted like soap. Lance Crackers were the worst for tasting like Dove bar soap.

And the worst soap was the orange Dial soap. This has to be the cheapest soap in existence, and it's the soap that we had to use in boot camp. The last thing we wanted to open was a care package that smelled like boot camp.

In late afternoon, we'd be called by platoon for evening chow. Sometimes, we actually had something fun to look forward to afterwards.

One of the officers had a laptop and projector for making presentations. Several people had gotten their hands on some DVDs. So a 7-ton with a quadcon on the back was set up to be used as a screen. It was put on a decline so it resembled theater seating. Some people brought chairs or boxes to sit on, but most of us sat on our poncho liners on the ground.

Before the movie, we had some beach balls we would hit around like we were laying out by the ocean. I tried to get there early because the speakers that were used were just ordinary office computer speakers. The picture was also somewhat small, so the earlier I got there the better I could see and hear.

After the movie was over, it was time for bed. Chairs and boxes were taken back to their original locations and we retired to our tents, hopeful the next day would bring good news.

At first, the movies were a great idea and helped keep morale as high as anyone might expect. It gave us some type of structure and something to look forward to on a daily basis. But as time passed some nights there would be no movie simply because there wasn't one to be found. Eventually, even movie night lost its luster.

People were beginning to get sick. The weather was hot and the temperature change was minimal. We guessed the physical sickness was our body's response to the emotional homesickness everyone was feeling.

Everyone's fuse was lit and emotional bombs were bursting daily. As tempers flared, fights broke out. It was best to stay away from as many people as possible. I even got in a fight with Ryan and cussed him out for being cheap and selfish.

I had to stop thinking of home. The more I thought about it, the crazier it made me. This meant I had to stop writing in my journal. It meant I had to stop writing letters home. This was a time for me to read my book and play the character of someone else in someone else's book.

I became a hermit and kept to myself. I stayed in my tent and read all day. If I went anywhere, I went by myself and tried to avoid all contact.

Then someone had the great idea of having classes where we could all reflect on what we had gone through. It was about the last thing any of us wanted to do, but we didn't have a choice in the matter.

We had a class on post traumatic stress disorder (PTSD) given by the Chaplain. What a joke! Giving this type of class to a bunch of male Marines was not the best idea. I mean, they were asking us to look for signs in our peers of emotional or psychological distress. Guys don't do that sort of thing—especially Marines! I've never heard so many homosexual jokes in my life. No one learned a thing in that class because we were all too busy cracking jokes to pay attention—much less retain anything.

As we were walking back afterwards, 7-tons with Marines in the back rolled past us. They had received word that they were next to go home, and away they went. If it wasn't sad enough for us to watch as they got to do what we wanted so badly to do, the whole unit sang us a song as they drove by ...

> *"Na Na Na Na, Na Na Na Na,*
> *Hey Hey Hey, Good-bye ...*
>
> *Na Na Na Na, Na Na Na Na,*
> *Hey Hey Hey, Good-bye ..."*

The next class was the one that sent me over the edge. It was a class on unsafe acts that we witnessed during the war and how to be safer next time. Are you kidding me? Normally this would have been as humorous as the PTSD class. But for some reason, this class actually sent me into a rage. In desperation, I had only my journal to turn to.

May 13, 2003. It seems like a lot of time has passed since I last wrote in or even opened this book. Although time may have passed nothing else has changed except for the sun waxing across the sky every day. Every day I get more and more stressed out emotionally. My mind has grown very weary of this place. I'm not even in the mood to write now, but what else am I going to do?

The weather has been outrageous. Yesterday the temperature was well above 100 degrees Fahrenheit and the heat index was above 130. And we're not allowed to dress for the weather. If we're up and about, we must be in cammy pants or coveralls. The only time PT shorts are authorized is if we're participating in physical training. I'll be damned if I go for a run when it's 130 degrees and there's no shower or air conditioning to return to.

We had a sandstorm the other day which usually cools things down a little but not this time. Along with the sand came a little rain, making for a very humid heat as if it wasn't already bad enough. And along with the rain came all these little critters in search of high and dry land—our living areas.

I've never had any fear of spiders before, but these camel spiders spook me a little. Seeing a sand colored spider bigger than a large rat seemingly out of control and rampaging toward you might spook most anyone.

But they aren't the only venomous creatures around here— scorpions are rooming with us as well. I've seen four over the past week. These little things don't look that scary, but just one sting can painfully end a life. That kind of power in such a small foe gives me a certain respect for them, knowing that death could be lurking right before my very eyes.

Then there are the flies, with their diet of filth, death and disease. As I lay here in my 2-man tent, I have about twenty or more scavenging my body and tent for food. I detest these disgusting creatures, especially the ones who insist that my face is the best place to land. It even gives me a feeling of claustrophobia if I think about it.

Sure, I could shoo them out of my tent and close myself in. But then I would suffocate in my new sauna. So what's worse? Flies? Heat? Death? In this matter, I believe they are all equal.

One day an idea struck me to improve my situation, or at least give some kind of relief. I went outside in the 130 degree sun in hopes of catching some rays. I climbed to the top of a quadcon on a 7-ton where I slept during the war. I had an iso-mat to lay on, a bottle of water to hydrate and pour on myself to keep cool, and some Skin-So-Soft oil to accelerate my tan. There was a good breeze up on top and few flies, so I felt more comfortable than I had in quite some time. I was going to lie on my back for an hour and then flip for another hour. But thirty minutes into this, Ryan climbed up and told me I had to get down by order of SSgt Lopez. He seems to have it in for me these days and I see a confrontation in the near future. I don't think I can hold my tongue much longer.

I feel strung out now and I'm not in the mood to do anything. I keep to myself these days. That is the only reason I haven't had any more outbursts than I have already had after cussing Ryan out for being cheap and selfish. Russ is a mean bully who gets his satisfaction by poking fun at others' weaknesses. And Ryan is right there with him, following like a lost puppy.

Then there is their " friend," Sgt Chavez. They talk shit about him constantly behind his back. Sounds like a great friendship to me. I wonder what they are saying about me right now. No, I don't wonder—it's a waste of my time. I would rather befriend these eight flies sitting here on my right leg. Look—two of them are wrestling right now. How cute.

A couple of days ago, I was hungry so I reached for my bag of goodies but it was not where I had left it. I immediately got up and began searching for it, but it was gone. I asked Ryan if he knew anything about it, and he gave an abrupt reply saying it must have been thrown away during a morning police call.

In that bag had been all the best food items from all my care packages along with other items distributed from large care packages. I had been living off them, and they were the few personal items I had. All of the nice pens Mom had sent me (the silver ones) were in there. And there were other items that were important to me. The loss was devastating, and to think that Ryan knows more than he spoke was beyond disappointing.

A few days ago, we were told we could make a phone call home. I wanted to wait until the 11th—Mother's Day—to call Mom. But when I tried to make my call, I was told it was too late. A few minutes afterwards, SSgt Little came by bitching that he had just then gotten cut off his call. If it had been too late, how did he get cut off? Well, that's an easy question—it's because some of our leaders are two-faced bastards who think they are above the common principles and traits of being a true leader. In other words, I work for hypocrites who demand that I do as they say, not as they do.

And now, to top it all, they make us attend a class on safety where they talk us into reflecting on unsafe acts during the war. Then they reprimand us for failing to correct those actions! This class was a big trap to punish people. They kept asking what we could have done to make things safer. Less than two months ago we were sent to war and into combat, and there we sat getting a class on unsafe acts. Well, I have a solution ... never enlist and escape the danger of stupidity!!!

Then there's the infamous quote you always hear ...
"It could always be worse."

No, no, I really don't think it could be.

Chapter Twenty-Five

Days later, we discovered why our Christmas had not come. Apparently there had been a major communications breakdown. Apparently there was a logistics officer somewhere in the States who had misplaced our unit, and thought we were already home.

Major combat operations were completed nearly three weeks ago. Our support role in the war was over. If we weren't in Kuwait, we must have been home. But we weren't home. How do you lose an entire unit? Were we that unimportant?

Once we were "found," we were soon slated to leave Iraq enroute to Camp Matilda. On May 21st, and only three hours late, we left An Diwaniyah on 7-ton trucks headed for a local airfield. There were only nine from my platoon who were able to go. Russ and I were lucky enough to be two of them.

We arrived at the airfield and prepared to board four CH-53 helicopters. There were ninety-eight of us total, and the birds could only carry twenty-four Marines each. That left two Marines stranded— SSgt Deller from MLM platoon with our company, and me.

Our choices were to stay there and wait for another bird, which could take hours or even days. Or, we could get our unit to come back and get us.

I could not have cared less if I never saw my platoon again, so that wasn't even an option. Plus, I wasn't going to pass up a chance to fly on a helicopter. So we sat down in the blazing sun beside our packs and waited.

It was only about an hour before our skin was burning in the sun. We had to get up and move under shelter. In the tent we relocated to were eight other Marines who were also waiting on a flight. We talked with them briefly about our experiences. For them, this was as far north as they had been.

One of the two female Marines was saying how sad she was and how much she missed her husband.

"Where are you from?" I inquired.

"Texas." she replied.

"So is your husband back in Texas waiting for you?" I asked.

"No, he's back at the air station in Kuwait." she said. "He's also a Marine."

"Oh. So how long have you two been apart?" I asked with some hesitation.

"Today makes three days and it will probably be a couple more before I get to see him again." She pulled out her mobile phone to call him.

"Ha!" I sneered in my smart-ass tone. I was actually very disturbed by this. I couldn't believe my eyes and ears. Why does she feel the need to even speak to me in the first place, and then to voice these concerns to me?

I've been without Abby for nearly four months and I've only been able to speak to her twice during that time. I walked away because I didn't want to get in a pissing match of, "Who misses who the most." But what I really wanted to say was, "Try spending four months away from him and only get to speak to him twice over that period on a monitored satellite phone with people waiting in line behind you for their turns as they listen to your entire conversation. Then be cool with the fact that you don't know when you're gong to be able to see him again. And be the one who has to tell him its okay when he's crying and begging for your return home."

And who carries a cell phone to war anyway?

About three hours after the first four birds took off, two more touched down. There was some confusion as to whether we were going to be able to take the flight, though, as none of the pilots knew anything about us. They didn't want to be responsible if there was an accountability issue.

At the last possible minute, we got the green light to go. I didn't have my gear ready, so I had to hurriedly get it together and put the pack on my back. As I made the two hundred yard dash to the bird, I almost passed out.

My pack must have weighed 150 lbs. It was also off balance, so it was awkward to carry. When these packs are out of symmetry or are loose, they bounce and fall all over the place. It felt like I was carrying a midget on my back as he was kicking and squirming to get away. On top of this, being in the desert had made me weak and the combination of the blazing sun and heat generated by the helicopter almost made a heat casualty out of me.

But I finally made it to the bird before it flew away, and collapsed into the last seat. This was actually the best seat in the house, aside from the pilots' seats. The ramp on the back was only raised halfway so the view was more than just a good view ... it was liberating!

The country we had just trudged through and taken over had actually taken us as prisoners in a psychological prison. I didn't realize it until this flight, when I was able to look down at it from a distance and no longer see all its monotonous colors and imperfections that had haunted me. I felt a peace I had not felt for a very, very long time. I closed my eyes and drifted off to sleep with a smile of hope.

I woke up as we were touching down at the Al Jaheed air base near Kuwait City. It was strange because we made a runway landing rather than on a helo-pad. I wasn't even aware that helos were capable of that but we did it. I looked down at my watch and noticed that nearly two hours had passed since we had taken off.

Some guys came to pick us up, and we rode north to Camp Matilda. The ride was so nice and uplifting. There was electricity, city and street lights, cleanliness and order. There were other cars on the road. There was life.

Wasn't this the same place we had so hated just two months earlier and couldn't wait to leave? After going through war and being in Iraq, this once-despised place now felt like the best place on earth.

We arrived back at Camp Matilda after a peaceful forty-five minute drive and could hardly recognize the place. There were more showers, Port-a-Johns, and a PX. The food in the chow hall was desirable, and soft drinks were now available. There were even Pizza Hut and Burger King stands. Our tents now had air conditioning, and best of all, there was mail waiting for us there.

I had four care packages of food, magazines, and hygiene items. Abby really hooked it up this time. She sent me a Zippo lighter that had an American flag on the front. On the back, it was engraved:

Carrying Your Love
With Me

Cpl. Eric J. Cox
2d AABn, H&S Co., Det D
Camp Matilda
Unit 76689
FPO AE 09509-6689
Love, Abby

Her gift was truly amazing. She knew that I collect Zippo lighters, but she couldn't have known my reason for doing so.

My grandfather, Papa Jones, passed away when I was not quite six years old. I have three photographic memories of him, and other memories of feelings.

I can remember sitting on the lounge chairs beside their swimming pool and talking about skeet shooting and the RC model airplanes he built and flew. I can remember the charisma about him that was so contagious, so ambitious, and yet so peaceful. I remember thinking that he was a man who could do anything, and I looked up to him as a role model.

I remember him smoking a pipe. He had a silver Zippo he used to light it. The Zippo was engraved with a flying duck, as he loved to hunt, along with his name, "Odell." The engraving as I remember was not your ordinary engraving. It was more like someone had carved into the lighter and painted the area beneath the surface a deep red. The silver and burgundy lighter was a piece of art.

When Papa was hospitalized with only few days left to live, he asked to see "his boys"—Darren, Stephen and me. At the time, the VA hospital did not allow visitation by young children, but Mom, my Aunt Lorraine and Grandma Jones "smuggled" us up a back stairway so we could visit him. I was nervous about what I was going to see, as I was too young to understand such a thing. But as he lay there, all I remember thinking was how content he was with his family by his side. It was just his time to go, and his serenity would forever preclude my having any fear of dying so long as I was loved.

Since then, I have always been infatuated with Zippo lighters. I can remember getting my first Zippo and taking it out and looking at the bottom of it. The logo was different than the logo on Papa's Zippo, so I was very disappointed. I thought I had a fake but then later learned that Zippo had changed its logo from the script style I remembered to the block style my new one had.

The only way to get the Zippo with the script style logo was to buy used lighters. Mom is an antique collector, as I believe most mothers are, so I always loved going antique shopping with her because I was always in search of Zippos.

But I don't discriminate when it comes to Zippos, be it old or new. My reason for this is that I hope to one day be like Papa Jones to my grandchildren and have quite the collection to pass down.

This Zippo that Abby has sent me gives me a sense of wisdom. I can just hear my grandkids finding it and saying, "This is the lighter that Papa Cox carried in the Iraqi war!"

Then I began to worry something might happen to it or that I would lose it. I can't believe that Abby really wants me to use this lighter. Is she crazy? I was getting stressed out just having it here. I couldn't think of a good safe place for it. I decided I'd just have to hold it until I figured something out.

To lift my spirits even higher, she also sent me two very sweet letters that were both from different times. One was sent back in February before she had even received a letter from me. The other was sent a month after the war was over ...

Eric —

Well, hello beautiful. Gosh, I hope you come home soon ...

How are you? I worry about you constantly! I watched this thing on TV about how the military tries to train you guys how to be "normal" again after you survived this war. And I personally don't think you can be "normal" again. You have lived through things most people will never see in a lifetime. Did you see horrible things? Do you want me to ask questions? You know me—I always have millions of questions. Should I not ask? I want to know only what you want to tell me, but I am curious about everything. I am curious about how you've changed. Three months of your life have gone by and I don't know anything! Like I said before, I wish I could have been there beside you through it all, but that's not how things have worked out. I hate, so much, that I have not been a part of your everyday existence, but I think I hate it more that you're there in a strange country and not where you want to be.

Are you O.K.? Is there anything I can do to help you? Should I try to help you? Or should I back off and mind my own business? I just love you and I want to be helpful and not pushy. Tell me whatever you think.

I look forward to the day I can hold you again. Every day without you I try to remember all the memories we have shared up until this point. I think a lot about that day when I first met your family—on Thanksgiving—and all the times you picked me up from work. I miss you more than anything I can put into words. You are my life, my everything. Hurry home.

Love always,
Abby

At 2300 that night, I was fast asleep and slept very well. I couldn't even remember the last time I had slept that good.

At 0700, Russ woke me; it was time for breakfast. I also couldn't remember the last time I had food that tasted so good. We both ate so much that when we were done it was hard to get up. We did nothing but relax and play Scrabble for the rest of the day.

The next four days would be just as relaxing, as we waited for the others in our platoon. We had beaten them back to Camp Matilda. They had to drive back from Iraq on broken down vehicles. Ryan was a bit upset to say the least when they finally did make it back.

Over the next week, it seemed as if we were living a life of luxury. We only worked a couple of hours a day, if that much, for packing up our parts and equipment. The remainder of the time was spent doing as we were known to do.

The PX was selling some kind of world cell phone for a couple hundred bucks or so. We thought about buying one just for the heck of it so we could call home anytime we wanted. But we decided we had made it this long, and could hold out for a few more days. It seemed like it would make our return home even more special that way.

This week and a half proved to be the most like Christmas because we knew we were really leaving this time. The window they had given us was the week of May 28 through June 3rd. As luck would have it, the first six days came and went but Christmas did arrive on Day Seven – June 3, 2003.

We finally made it to the plane in Kuwait City on that day and knew it was only a matter of hours before we would be back in the States and see our families again. We began boarding, hoping to have the same seating arrangements we had enjoyed on the trip over, but really weren't all that concerned about it. They called for all Staff NCOs to come up front and all the NCOs and troops were to ride in coach.

We knew if we boarded the plane first we would have to sit in the very back, so we played the "after you" game. Our goal was to board the plane last in an attempt to be pushed up into business class or to be close enough to business class that we could make the move if the opportunity presented itself. As we sat in the very front of coach class, we noticed there were seats available in business so it was only a matter of time before we would make our move. But then the flight attendants came back and said there were seats available up front if anyone wanted them. Russ, Ryan and I were the first ones there.

Our first stop was Rome, Italy. We had to refuel there before making the flight to JFK International in New York. When our plane was on the ground in Rome, they attached the air stairs to it to give us a place to smoke outside the plane itself. But they wouldn't allow us to get off the stairs to step foot on Italian soil. We believed the reason for this was that Italy didn't want to give any appearance of aiding the US in the war.

Well, Ryan, Russ and I had never been to Italy, and we wanted to be able to say we had. In order to do that, we had to stand on the ground. So there we were at the bottom of the staircase, trying to sneak an extra step off.

Two security guards stood in our way. They were dressed in all-black custom tailored Italian suits. One was wearing a Gucci watch unlike anything I had ever seen.

I was wearing a green five-dollar watch Mom had given me before I left. I jokingly asked him if he would like to trade with me in an attempt to befriend him into allowing us to step off the staircase. He smiled but said nothing.

Then we learned that somehow, someone had worked a deal to stock our plane with beer—enough for everyone to have two. When the car carrying the beer pulled up to the plane, the two security guards knew help was needed to unload and load the beer. So Ryan, Russ and I were allowed not only to step foot onto Italy—we assisted with the first beer run we had made in nearly four months! It was a three-fold victory for us—we beat the system, went to Italy, and had beer!

Back on board and enroute to New York, the attendants handed out our two beers each. Even those who were under the age of twenty-one were allowed to drink. Everyone was smiling, laughing, making jokes and having a great time. But since not everyone had wanted their two beers, some were left over. After beer #4, with our reduced level of tolerance, we passed out for a couple of hours until we neared the States.

We woke to the sound of the pilot's voice coming over the loudspeaker announcing, "We have just crossed the line into New York. We are now in the United States of America."

The plane went crazy! Everyone was cheering, clapping, giving high five's and anything else they could do to show their joy. It was such a good feeling to be back in our homeland. Flying in and touching down at JFK was only a blur, as thoughts of home were the only visions we had.

AT JFK, we had exactly an hour to do as we pleased before we had to be back on the plane. We wanted to make calls to let our families know how soon we would be home. And we wanted to find a bar.

We ran searching through the airport. Finding a bar seemed to take forever. When we got to the first one, we found all the officers including the Colonel were already there; we were the first enlisted.

We ordered a beer and a shot of Tequila and toasted to being back in the United States. Civilians were coming up to us and thanking us for our service. Russ looked at me and said how proud he was for the first time since all of this began.

I ran out of the bar and made a call to Mom to let her know we were in New York and that we would be back in Cherry Point very soon and would then be bused back to Camp Lejeune. Her excitement was uncontrollable and then I was back in the bar for another beer.

The plane ride from JFK to Cherry Point was silent; most everyone was in the zone as we thought about seeing our loved ones again.

We exited the plane at Cherry Point and collected our bags. The long walk to the buses seemed like a walk in the park as we knew what would happen next.

Communication on the bus was mid-level; no one really wanted to engage in deep conversation. There were just too many other things on our minds.

As we neared Camp Lejeune, we could see "Welcome Home" banners everywhere that families had made using sheets and paint with words and images of love and support. We pulled into Courthouse Bay, and could see people waving and cheering. We had one turn to make before we would be in the parking lot of our barracks, but the bus passed right by it. We could see our families but didn't even slow down for them. We had to go to the armory first and turn our weapons in before they could let us loose. Luckily, we didn't have to clean the weapons before turning them in, or we would have been there for hours and it was already growing dark.

After a few minutes, we got back on the buses and drove to the barracks. We made the final turn and could see people running towards the parking lot. We curved around behind the barracks and emerged into view. Everyone was there cheering for us as we stopped in almost the very spot we had departed from those four long months ago.

I scanned the area looking for my family. I could see Abby, and I could see Mom behind her. I could see Dad and my brother, David. My sister, Kim, was there with her three children—Josh, Julia and Lindsey. They had all come to welcome me home.

I got off the bus, and this time it was just me. No weapons; no packs; no weight of the world upon my shoulders. Abby ran up and hugged me and we held each other close for as long as both Mom and I could stand it. Mom cut in and we hugged. Kim and I hugged. I knelt down and hugged Julia and Lindsey. I turned to Josh and we hugged. David and I hugged. And lastly, Dad looked at me with pride and hugged me without hesitation. It was good to be home.

As we gathered our bags and prepared to leave with our families, I couldn't help overhearing another Marine talking to his wife. Even though he was speaking only to her, he exclaimed as if he was making an announcement on behalf of everyone ...
"Get me out of here—I don't ever want to see these guys again!"

Well said.

FROM THE AUTHOR

My participation in the Iraq war ended on that day in June, 2003. CPL COX is my answer to the question, "What was it like?" It was written in part to help me cope with my own issues, but may also provide others with a certain degree of insight and understanding into the complicated and extreme physical, emotional and psychological pressures that sometimes accompany life in the military. Although it is my story and I am fully aware there are many others whose experiences were very different, I am certain I am not the only one who can relate.

As my memoir ends and this particular chapter of my military service is complete, there are a number of unanswered questions. The happy ending my loved ones and I had hoped for would be far, far away. The full story had only just begun. Upon my return home on that beautiful evening, I had no idea what I had started to feel or what I was about to go through. I had no idea as to the magnitude my life had been changed by the war, or the relationship issues, rising inner turmoil and rage that would result.

Perhaps one day there will be another book that will answer the questions and bridge the gaps. But next time, the title will have to be different ...

Semper Fi,

Eric J Cox
Pvt / USMC

LETTERS FROM HOME

THE COURIER-TRIBUNE, ASHEBORO, N.C., Wednesday, March 26, 2003

Local residents serving the U.S. in military

Editor's Note: *With the war beginning in Iraq,* **The Courier-Tribune** *is asking families to send us information about loved ones who are in the military.*

Keep the information concise. Include the person's name, what unit he or she is serving with, parents' names and city of residence, high school and year of graduation, where they're stationed, if you know, and an email or regular address where people can send them letters. If you have a photo, send that as well. We'll publish these as often as possible throughout the duration of the war.

In addition, if you'd like to share any emails from your loved one overseas, send those as well. We will publish as many as space allows.

Email them to ajordan@couriertribune.com.

For more information, contact News Editor Annette Jordan at the above address or by calling (336) 626-6140.

■ **Cpl. Eric J. Cox** is serving with the U.S. Marines 2nd Amphibious Assault Unit. The 1999 Asheboro High School graduate is the son of Jimmie and Cynthia Cox of High Point and the grandson of Emma Jean Jones and W.R. and Mavine Cox, **Cox** all of Asheboro. His address is Cpl. Eric J. Cox,

2nd AABN H&S Co. Det. D, Unit 76689, FPO AE 09509-6689.

■ **Recruit Jonathan E. Sypole** is currently in boot camp. He is a 2003 graduate of Asheboro High School. His mailing address is RCT Sypole, Jonathan E., PLT 3040 3rd BN L CO, BOX 13040, Parris Island, SC 29905-3040. He is the son of the Rev. Jeffrey and Laurie Sypole of Asheboro.

■ **Lance Corporal Jeffrey H. Sypole Jr.** is currently stationed in Iwakuni, Japan. He is a 1999 graduate of North Davidson High School. His email address is jsypie@hotmail.com and his mailing address is: LCpl Jeffrey Sypole USMC, PSC 561, Box 1087, FPO AP 96310-0021. He is the son of the Rev. Jeffrey and Laurie Sypole of Asheboro.

■ **SPC Joshua Bowland** is serving with the 1454th Transportation Co. A 2002 graduate of Southwestern Randolph High School, he has been with the National Guard **Bowland** for over a year.

■ **SPC Brandon R. Ellis,** in Army Reserves and deployed to Kuwait, is driving a Humvee with the 3rd Infantry Division out of Ft. Stewart, Ga. A 2001 graduate of Asheboro High School, he was attending Randolph Community College in automotive systems

technology. His parents are Barry and Rita Ellis of Asheboro. His fiancee, Ashley White, lives in Asheboro. He can be reached at SPC Brandon Ellis, 422 CABN, Unit 93800, APO-AE **Ellis** 09303-3800.

■ **Lance Crp. Michael Oakes** serves with the U.S. Marines as an electrical engineer on the EA6B Prowler with the VMAZ-Z unit stationed in Cherry Point. He is currently deployed to Prince Sultan Airbase in Saudi Arabia. He is married to Britten Nichole Oakes of Asheboro **Oakes** and is the son of Debbie Thomas of Asheboro and Mike Oakes of King. He can be reached at LCPL Oakes, VMAQ-Z/USMC, 363 AEW/Box 25, APO AE 09882.

■ **Sgt. Jason E. Murray** is serving with the U.S. Army, in E Battery, 5th Battalion, 52nd Air Defense Artillery (Patriot Missiles). He is the son of James and Carolyn Murray of Asheboro. A 1996 Asheboro High School graduate, he is now in Camp New Jersey, Northern Kuwait. His mailing address **Murray** (not allowed email at this time) is Sgt. Jason E Murray, E Brty, 5/52 ADA, Camp New Jersey, APO AE 09329.

Dear Eric J. cox,
How are you doing in war right
know? I am not in war but I am
doing okay. Do you like being war
right? If I was in war I would want
to go home. I just can not think of
getting shot. It would probably
hurt if I did or you did. You
must miss your parents? Don't you?
If was there in war without
seeing my parents for long time
and worring that I might never see
them again, I would be crying
all night long. I hope you
make it and come home to
your parents. They probably thinking
about you right now and crying
and preying for that to happen.
And I am hear Supporting
the war.

I am Micheal Rosas in Asheboro
middle school in 6th grade.

SAMS Middle school

Sincerly,
        Michel Rosas

Dear Cpl. Eric J. Cox,
   I hope that you guys are doing fine. I wish we can help in anyway. Please be very careful when you are in the battlefeild. I will be praying for. The Idan Kember that God is with you. God Bless America, and the Troop! Pease write back!

From,
Julie Olivares

Dear Eric J. Cox,
   I'am glad that you are serving for your country and that you are going to beat Iraq in war, But I want to wish everyone that is there at war good luck and be raseful.
   And are you worried that maybe you might have to lose a friend or family the war, if I was you right now I wouldn't like to go to war, so thank you for keeping Iraq from killing everyone so there for everyone says thank you.

Sincerly,
Daniel CKing

Dear Eric Cox

Are you and the other people
board out there will Now you
whont because you will be
reading. So you whant to
Asheboro High School Pid you
go all year there. Is it scarcy
out there? You must be braue
to go out and fight for your
contwry and we thank you for
that. Are you bord some time!
Do you go out and fight or
shot or do you help other peple
or do you ride a Airpane and
shot bumps down or what
do you do. So you in the
U.S. Morfnes like the carmuson
t.v. like that one or something
Have you talk to one of your
lovedones or relttves or
something and tola them what
going on or something. Do
you see people get shot
or any of the Iaqu people
or what you call them, what

else do you do out ther.
Our teacher tell some of the
thing that happen and our Momare
Dad tell us what goes on.
Well I geose I will let you
go on I hope you Injoy
reading my and other people
letter thank for reading

Sin.
Brittany Hooker

Jesica

Dear CPl. Eric J. Cox,

How is everything going down there?
Everything is the same here in Ashebaro.
I am Just letting you know that I care
about you being in the war to save hurdes
of peoples lives right now. Are schools
case's about each and one of you down
there, be shor to tell them that. Ute Family.
is doing Just fin. Got to Go "bye."
                              P.S. write beck

                    Sensedly yoors
                          Jessica
                             Euan
                          6th grade south
                          Ashebaro middle
                          school NC.

Dear Cpl. Cox:

I got your address in the Asheboro Courier-Tribune today and I wanted to drop you a line to give you my support and to say thank you for what you are doing to protect us and to keep us free.

My two sons are also Marines. Jeffrey is in Iwakuni, Japan. He is your age. He graduated from North Davidson in 1999 and joined the Marines right away. He is supposed to get out in September, but it doesn't look good for that right now! Jonathan is a graduate of Asheboro High School. He is actually a 2003 graduate but took early graduation and is currently at Parris Island (I'm sure you remember those days!)

I am the principal at Ramseur Elementary School and my husband is the pastor at Brower's Chapel United Methodist Church in Asheboro. I just want you to know that we are praying for you and all our men and women who are serving their country. May God bless you and keep you safe.

Sincerely,

Laurie Sypole

3-25-03
Tuesday

**Praying you will
be uplifted today.
You are in God's care!**

Eric,
I wanted to send a note to let you know we are thinking about you. I hope everything is going OK. I can't wait to hear you are back at home! Come over and I will make you all of the "Cow Pattie" cookies you can eat. I talked to your parents Sunday week - they grandmother last. They are very concerned about you, Please be careful. →

Katie is doing good, she is also worried about you. We miss seeing you, you are a special person! Is there anything you need that I can mail to you? I understand it takes the mail 17 days or so. If you need stamps paper, candy or whatever I would be happy to send it. We have been watching the War on TV and it is hard to understand that it is real. I am not a good letter writer, but I think of you often and you are in my prayers. The rest of the family is also thinking about you. Randall is jumping every night and Mindy is always doing homework. Katie does not get home much. I miss her too. Take care and I will write later.

Love
Cindy

3·20·03

Eric ⭐⭐⭐

Hi!!! I've been Thinking about you non-stop
since 7:15 last night when Bush annouced that we are
at war. I'm not sure where you are rightnow... I'm
assuming you crossed Kuwait's borders and into Iraq
It will probably be harder for you to write from now on
so I figured I would write to you again instead of
waiting for a response.

It's kind of interesting because I get to watch
it all on T.V. I just watched This special on MTV called
"Gideon in Kuwait" I don't know if you are familiar with
him but he is an MTV news reporter. I Thought it was
cool to watch because he stays at Grizzly Range w/ the
5th marines. Most of the other reports come from about
50 yards away from The base on T.V. but The MTV report
showed it all up close, talked to all the marines and showed
Their living quarters, food, everyday life ... etc. So
now I don't have to use my imagination when you write
me about living in tents w/ little or no hot water. Its
weird because I wake up and turn on The T.V. and
it's dark in IRAQ. & Practically every channel is
showing coverage on the war.

But enough about That... How are you besides
the circumstances? I hope all is well. I'm just chillin'
over here in Cali working and going to school. OK
well barely going to school... my classes are lame.
I am taking Italian now, That's pretty cool but
everything else is boring. I snowboarded at Bear
Mountain last weekend... I haven't been there in
sooo long. I Thought about you while I rode the
Bear Mtn. Express (the lift we were on) unfortunately

It rained... not snowed on Sat. so I only got one day in.

Work is good at the Bakery The Mexican cooks and Dishwashers don't speak very good English and I don't speak very much Spanish so communication is minimal. They like to make fun of my white girl CA accent and ask my how I say things & talk about me in Spanish But its cool because me & the other white girls make fun of them in English and they have no idea what we are saying.

So anyways... Where are you? Have you met any resistance? They say on the news that in Southern IRaq the iraqi troops are surrendering by the hundreds and that 2 major/key airfeilds have been taken by the coalition in western Iraq. Sounds succesful

OK soooo, don't think I'm crazy but I talked to your mom this morning No I'm not stalking you don't get your hopes up 4/K. But she wants me to send you calling cards so you can occasionally let (she keeps hope because) her know you are ok. She said "Eric is like Teflon, trouble just rolls of him". She is so sweet, we shared some tears & some laughter about you but she is staying strong.

So anyway I hope you are safe and doing ok. Write when you get a chance. You are in my thought. Come home soon.

always Morgan

P.S. let me know if I can send you anything... disposable cameras, necessities, magazines, etc ok?

Sunday Afternoon
3/23/03

Dear Eric,

Well today has been our worst day
since the day you left and I have
a feeling this may also be true of you.
We have watched all the TV news for
the past 4 days since the war started
and to a small degree have lived it
with you. But today the reports are
about 12 marines being killed by
Iraqis at Al Nasaria (can't spell it)
when they pretended to first be giving
up to the marines. Also, other military
were killed and taken prisoner from
a maintenance convoy somewhere
outside and north of Basra. Later in
the day we learned that they were
army. The fighting in and around these
two cities seem to be the fiercest of
all for the ground troops. We can only
imagine that this is you and our
concern is very great.

Today I went to church with
David. I took him your letter
which we received on Saturday
only with your post card and ⌐

your letter to Ken. We took Kerri's
letter to her last night. Obviously they
were both thrilled to hear from you.

On Friday afternoon Abby came to
High Point to visit with Cynthia and
bring her a gift from her cruise. She
was still here when I got home so
the three (3) of us went to the Liberty
Pub for supper. As you can imagine
we spent most of the time talking
about you and had as good a time
as possible without you being here.
Zula came with her and we all
played with her - Abby had just had
her hair cut and of course she was
(Zula's hair)
a little doll.

We still have not sold the truck
or jeep. Today David is taking digital
pictures of both and it will put
them in the Trading Post. Maybe that
will speed things up.

Daddy went to the hospital and
stayed 2 days for fluid buildup.
He is back home now and doing

O.K.

Well this weekend coming up is when we are to move out for the furniture market. Your mom is staying with Kim and I will be at Kim's or David's. I probably will do some of both to avoid the longer drive every day.

Today Kurt Busch won the Bristol Nascar race and of course Tiger won the Bay Hill Invitational tournament for the 4th consecutive time. The last time anyone won the same tournament 4 times in a row was in the 1930's. He has won 3 times already this year and has only played 4 times (he also won by 11 strokes).

Also, today we learned of the 101st airborne soldier who through 3 grenades in his own officers' command tent. One was killed and 15 hurt (4 others were serious). He was an islamic American but of course no one yet knows why

he would ever do such a thing. Be
sure to watch your back side.

its really hard to know how
the war is going. We know you're
moving fast and the predictions
are that allied forces will reach
Bagdad in a couple days. There
are still 3 Republican Guard
Divisions around Bagdad according
to our news. No one seems to know
for sure whether Sadam is alive
or dead by do know that he was
at least hurt the first night of
the war when his palace was
bombed. I know that you still
have a lot ahead of you. As a
father I wished I could give you
some advice, but in this case
you are the expert.

It will be a happy day for us
all when you get back safe. Please

tell Russ and Ryan that our prayers are with them as well.

I hope our letters are still getting through but realize that for a while they are probably not.

Please know the pride we feel for you being there. I never miss a chance to tell people and talk about you.

Know that I love you and write when you can.

Love always,
Daddy

Monday March 17

Eric-

Hello beautiful! I just wanted you to know how sweet your parents have been. Your mom talks to me every time I call her and she never makes me feel like she doesn't want to talk. Your mom actually makes me feel a lot better about you being gone than my own mother. I don't even talk to my mom about you very much because she makes me more insecure about our relationship. I bought your mom a little glass box while I was in Cozumel- I thought it could make her feel a little bit better. Or at least get her mind off you being gone. I just wanted you to know that you have great parents.

So I've been watching the news and it says that war is going to happen soon. and I want you to be careful.

your mom & dad called me last night and asked if you had called me. They thought you had called or tried to call but they weren't home. They both were so disappointed that they may have missed your call.

you said in your last letter that you did not want to call but I am sure glad that you did call. I was kind of offended that you did not want to call but I guess you have your reasons. Just don't shut me out! Its hard- I know but by shutting me out its not going to be easier.

I am glad that you are writing in your journal. I bought myself a journal but I haven't used it. Every time I start writing it makes me feel better that you will be reading it so they all end up being letters instead of journal entries. I hope you

don't mind; I just don't want you to think I
send too many letters. I just love you and I need
you to know whats going on.

Today you have been 38 days - it sucks. Its only
been a month & eight days but it feels like an
eternity when I am away from you.

I want to go back on vacation. Because it doesn't
feel right being here without you. Everywhere I go
reminds me of something that has happened with you,
and it makes me cry. At least while I was on vacation
I didn't have to think about all the times you called
me or text messaged while I was ~~somewhere~~ somewhere.
I just miss you so much and I don't like being
anywhere without you!!

Love you & miss you Always

abby

Hurry Home

Tues, Mar 25

My sweet Eric,

There's a website for your Batallion and it has pictures on it of some of you! The detail isn't good for the group pictures so I can't tell if it's you or not, but I'm going to pretend it is! I can at least see what things looked like while you were in Kuwait. There was one shot of a sunrise - if you weren't there to fight a war, I guess you could see some beauty in it.

I cannot begin to tell you what it's like to be trying to watch the news coverage of the war in "real time" and not know where you are and if you're OK. I keep trying to find a word to describe it, but I can't. You're probably the same - I guess we know now why it's simply put - "War is hell." I've had to turn it off for a while - for the first 3 days I watched it pretty much 24 hours a day. Instead, I've put out more pictures of you in every room of the house - pictures of your beautiful smile - and try to think of all the good times.

So many people ask about you - Sam Coble called yesterday and Cindy (Katie's mom) called Sunday night. Abby called again last night. She had gotten a strange call and thought it might have been you trying to reach her. She was having a very tough time.

We leave Friday for the furniture market ordeal. Sunday, when I had to turn the television off, I was absolutely furious with Jimmie for making me rent the house. (I was also furious with myself for not forbidding you to join the Marines!) But I guess you probably won't be able to call anyway - I just pray that if you should somehow get a chance you will call my cell phone or Kim's house. Maybe having the birds around will keep my mind off worrying about you a little.

Guess who came to see me yesterday? Miss Priss!! I was so tickled to see her - I gave her milk and food so maybe she will start "visiting" again. I want to kidnap her (catnap?) and take her to Kim's.

I had some news today from Elaine - the woman that rents our Asheboro house - she is apparently very close to leaving her husband and moving back to Tennessee. Which really seemed like good news to me - maybe it will force Jimmie to let us move back. I would move back tomorrow if we could - I really want to be back in Asheboro when you come home.

We could not be more proud of you, Eric, for what you are doing, or admire you more. We also could not be missing you more, and looking forward to this being over and home again.

All my love,
Mom

(Thought you might find the "diamond" article interesting - just in case one of us..?)

Dear Cpl. Eric,   March 26, 2003
    I saw your picture in the Courier-Tribune. What a great thing you are doing for our nation! I am with the youth group at First Baptist Church in Franklinville.
    Anyway... My name is Sarah. I am in the 6th grade and I am interested in gymnastics. I appreciate all that you are doing for our country!
    Please, don't give up hope! May God be with you and your unit. Tell everyone at your base that everyone back in the United States prays for you all every day and we think about ya'll all the time!
    You and all the troops are so brave to fight for our country and risk your lives to help make the people of Iraq free.
            God bless you!
It would be great if you responded so I could have better understanding of how it really is in times of war.

                        Semper Fidelis,
                        Your friend,
                        Sarah

3-27-03

Cpl. Eric J. Cox

Your picture & address was in the Courier Tribune on Wed 3-26.

We just want you to know you & all your group are in our prayers. Thank you for what you do. God be with you at all times.

Our grandson (Travis - 13 yr) help'd tie yellow ribbon on tree at Randleman Middle School on Mon. The tree will stay up until you come home.

Take care!
Carol & Dink Routh

HANG IN THERE!

Erick-
Please know that we think of you everyday and are following the News daily. You have a lot of love and prayers coming your way from Asheboro, NC! We can imagine things are real tough where you are, but we know you guys know what you're doing! Sam and Sansra

Please let others in your group know how proud we are of you. We are in awe of your bravery and we support your mission! XXOO!

April 3, 2003

Dear Cpl. Cox,

I am a Spanish teacher at a middle school in Biscoe, NC and several of my students expressed a desire to write to some servicemen. I am from Asheboro and I saw your picture with address in the paper and I got the idea to let my students express their support for you and your fellow servicemen. We are praying for your safe and speedy return home. I did not do any editing of the students' work – it's all original work. ☺ I hope you enjoy these letters from "home." Take care and come home safe and sound.

Sincerely,

Lee Anne Woodall and East Middle School Spanish students

Biscoe, NC 27209

Amber Brady	Jessica Lopez
Kortnie Wall	Maritza Gallardo
Maira Aviles	Itzel B.C.
Matthew Stewart	Talissa Colquitt
Alexis Moore	Brandon Stohn
T-Jay Barrett	Oprah Graham
Kayla Smith	

Hi Eric—

Hope you are doing well & thinking of ya'll! Stay safe! Watch the bullets!

⭐ Jen

DON'T EAT PORK! ☺

ABBY'S SEXY ASS IS WAITING FOR YOU, SO HURRY HOME! ☺

CPL Eric J. Cox
United States Marines
2nd Amphibious Assault Unit
Camp Lejune, North Carolina

Dear Eric,

It is Friday, March 21, 2003. Today, as in the last couple days, we have watched the missiles and bombs rain down on Baghdad and fighting in many other areas of Iraq. Yesterday, the children in Julia's kindergarten class (at Farmer Elementary School) wrote letters and drew pictures to show their support for America, you, and the ones around you (Ryan and Russ are the only ones I know by name.) We discussed how brave you all must be to be a part of such an undertaking. They are very proud little Americans and very concerned for all of you're safety. After their letters were complete, the children stood (with Julia leading) and sang "God Bless America", and "I'm Proud to be an American". The teachers and I were overwhelmed with pride and emotion. You would have been very touched by the scene. All of you are supported by so many. We are sending you some goodies that I hope you will be receiving soon. I hope you enjoy their letters, pictures, and future treats. Please share with the Marines around you. When you get a chance, I am sure that the kids would love to hear from you and the guys. I believe that it is important that they have a personal connection to the events taking place.

Send any letters directly to:

> *Attn: Mrs. Beane's Class*
> *Farmer Elementary School*
> *3557 Grange Hall Road*
> *Asheboro, NC 27205*

Please know that as our minds wait and wonder of the events yet to come. Our hearts and prayers are always with you and the exceptional men and women serving along side of you. Please stay safe and we look so forward to your return. I love you and miss you much more than you can imagine.

Love,

Kim, family, and Mrs. Beane's Kindergarten class

March 26, 2003

Cpl. Eric J. Cox
2nd AABN H&S Co.
Det. D,
Unit 76689
FPO AE 09509-6689

Dear Eric,

    Forty five years ago I was a Corporal in the Marines. Today, I am retired at 65, and feel the best thing to do was to write to some of you guys. I sent an E-Mail to President Bush, in support of the United States, and our great Armed Forces. I imagine he feels alone at times, but probably not as much as you guys.

    Eric, I want to thank you for being a great American, fighting for freedom. As a boy, I well remember WWII, and the threat of our freedoms. No one wants war, but any good American must be willing to fight for freedom.

    About Asheboro, it is beautiful here, and is spring time. I have several friends, who might have gone to school with you. Melissa School Craft, who I've known since she was an infant. Melissa is at UNC-C, and is a Sophmore. Mary Williams is at Clemson. My wife June and I took Mary and Melissa to cheerleading finals when they were young teen agers. Rebecah Bunch is a Mars Hill. I've known Rebecah, for about 12 years.. She asked me to be her mentor, last year, for here senior project. I taught her to play electric bass. Lisa Morton lives two doors down from me. She is studying nursing and Randolph Tech.

    The Army will take care of you, all of your life. After my Honorable Discharge from the Marines, I had access to the GI bill. I used it to get a Masters Degree. Later, it paid for my Commarcial, and Instrument Pilot Ratings. One time, I found myself with no money to buy a home. I was able to get a Veteran's loan to buy a new home.

    Whate aver happpens, Eric, God is with you, as well as your family and friends. You are not alone.

God Bless you, Eric

Jim Davis

4-4-03

Greetings Capt. Erie Cox

Just want you to know that we pray for all of you each day. We are very thankful that you are risking your life for us and our country. Ernest, my husband, served 12 years in service for our country and my grandson is in the marines at an island off Japan. I am so proud and thankful for each of the men & women in our Armed Forces.

I saw your picture in last Sunday's Courier and recognized you as being one of the students at Asheboro High School while I worked there in the cafeteria. I worked at the booth in the corner selling pizza, subs-chick-fillet & etc.

Capt Erie we will continue to pray for God to watch over and protect you and to bring you all home soon.

yours truly
Ernest & Mane Rugg
Asheboro NC 27205

Mar. 28, 03

Dear Eric,

I'm sure you probably won't
remember me but I do you. I taught
(teacher asst) at Lindley Park for 25
years and I can't remember if you
were in my room or if I just
remember you being at Lindley.
I have known your parents for a
very long time, as a matter of fact
we saw them not too long ago
in Dixie Restaurant and they said
you were in Kuwait, also your
Grandmother Jones is a "gardening"
friend of mine, so I feel like
I know you also.

I saw your picture in the
Courier this week and just
wanted to write and tell you
that I pray for your safety
each night and are so proud
of you and all our troops for
the sacrifice you make every day.

Please know that all
Americans are supporting each

of you whether they approve
of the war or not.

We watch the news daily
(not just the 6 o'clock news) to
keep in touch with you boys.
My heart aches for your family
at this time but I know we
will __win__ this war and you
boys will be home soon.

Be very careful and take
good care of yourself and know
we in Asheboro love you very
much.

God bless you,
Evelyn Yow

PS
We live right off 42 not far
from Mrs Jones and I think
of you everytime we go by
her house, she is so crazy
about you grandchildren.

P.S.S. Eric I just got off the phone to Mrs.
Jones & she said it was David that
went to Lindley so I had the two
of you mixed up but I do
remember you as a little boy

Doris Beeson
Clarine Caudle
Mary Underwood
Bonnie Caudle
Roderon McCorquodale
Fannie Hunger
Jennie Wright
Eunice Johnson
Gladys Brown
Gladys Hudson
Marilyn H. Parri
Carol Routh
Betty Routh
Emme Routh

Buddy & Thelma Capps
Jack & Hazel Cobb
Sam Routh
Fred Hudson
George Brown
Fannie Poupore
Betty Pugh
Vernice Pugh
Ida Mae Stout
Ronald & Edith Little

You're in God's thoughts,
and in His heart–
a very special place..
showered with His love for you
and covered by His grace.

Hope Your Day
is Blessed -
With His Love.

Our Love & Prayers,
The Willing Workers S.S. Class
of Bethany United Methodist Church,
Franklinville.

"How precious also are Thy thoughts unto me,
O God! How great is the sum of them!"
PSALM 139:17 KJV

Eric,

Greetings! My name is Lauren Kilby and I am from Asheboro, NC, although now I am in Indiana attending college.

Recently, I was home on Springbreak and came across your and a few other peoples pictures in the paper, so I thought I'd write each of you and let you know I was praying for you! I know when I'm away from home I love it when I recieve mail!

I know its difficult being away from your family and friends, but especially during war I'm sure its even harder. Just remember that if you know the Lord, He will never leave you or forsake you!

I hope the sun is shining where you are and that you have a fabulous day! Thanks for all that you're doing for us!

                              - Lauren Kilby

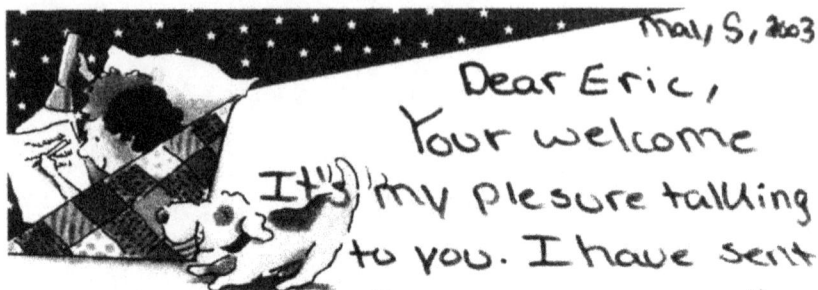

May, 5, 2003

Dear Eric,
Your welcome
It's my plesure talking
to you. I have sent
you a piture of me so you won't
forget me! and I hope you get
home soon, and yes sams mildle
school really like it if you come
and talk about what you did
in the war. The book,
<u>The Lord of The Rings</u> is a
great movie! I also hate
readind to. Oh and I'am Doing
good in school AB aroll I'am
looking ford to 7th grade! Just
to tell you again everything
here is doing fin. And I'll
tell the teach's that you
said "Hi" to.   P.S. write back
your penpail

Love. Jessica
Evans
"<u>Bye!</u>"

Dearest Eric,

How surprised I was to see your picture and to read of you in the <u>Courier-Tribune</u>.

I think of you often and I am so glad to have an address.

Thank you for what you are doing. Please know that you are in my thoughts and prayers.

I look forward to having you stateside very soon. Stay in touch as you can.

Love,
Mrs. Martin
3-30-03

April 9, 2003

Dear Eric,

I was delighted to see your photo and address in the "Courier - Tribune". I meant to write sooner, but several things have happened to delay my objective.

Barrett had to have emergency surgery mid - march at New Hanover Memorial. Another "stone" case. David and I drove down on Tues. Left @ 4:30, arrived at 8:15. They were waiting on us before they wheeled him into operating room. He had been very sick throwing up. and in excruciating pain for at least 48 hours. They operated and left a stent in for safety. They finally removed it - ouch. Barrett hates these things. It could be worse - he could be in Bagdad huh?

How are you doing? Our thoughts and prayers are with you each moment.

Who would have ever thought
that kid in my backseat
head bangin all the way
up I-95 to the "Warped Tour"
would, be defending our
Country?

I am proud to know you
and admire you for joining
the Marines. You have
grown into a handsome
young man. I hope I
get the opportunity to
see you when you
return home.

"Rudy" says "arrf" hello –
We Do now have her.
Barrett's sissy roommate
Tim did not like RUDY
and we had to adopt her.
What an ordeal. She misses
her buddies who hung out
with Barrett. Jumping the
fences and chasing the
squirrels, tearing out the plumbing
and "pooping + peeing" all ove the
house on Rose Ave. Ha! What a
wild dog. We have a line
strung in the backyard for
her to run on. She likes to
sit on top of her dog house and bark

I wish we could let her run but she'd never come home. She has escaped a couple of times "during the ice storms". Whew _____

I sprained my ankle at a "Swing Dance" workshop a couple of weeks ago — ouch! I was on crutches for a week and it was immobolized. Trivial to what you and the other soldiers are enduring in Iraq I know — but still traumatizing for a "Dance Teacher".

Well Buddy — I hope the next time I see you, you are lying on Barrett's' couch covered up under his infamous baby blue blanket, getting some well deserved sleep!

May God be with you and protect you from the enemy.

Always ——

Your "Warped" Moma

Shelly

Eric,

Whats up dawg. I have been thinking about you, and what your doing. I bet things are crazy over there. You will have to write me sometime or call when you get a chance. Tell me how you are + whats the plans for the future.

My parents gave me the clipping out of the Newspaper of you + your address, you look bad man. My parents are up to the same old stuff. They bought a place at Myrtle Beach. When you get back we should take a group of guys + dates down for a week. I see your folks are living in HP now, How are they? I saw Nance @ the Warehouse the other day, he's working there now. He's so funny. He could be writing novels + painting masterpieces, but he chooses to pack blue jeans and shoot the shit with Craig + Coy. Coy's crazy Remember these "Bumping any fuzz" or "Getting any strang lately". You will have to stop by the warehouse + say hello to everyone. It wont be long + I'll be there everyday.

I am writing this during a World Religion Lecture. My teacher just asked a question and out of my half listening I answered correctly. Way to go Me.

Me and my girlfriend Jenna, bought a dog together the other day; We named her Mya McKenzie. She's half lab + half Rottweiler, + mostly spoiled rotten. And I wanted to tell you I'm running Track for H.P.U. I walked on to the team. It's funny, I am the only White guy on the Sprint Team. My races are the 400m (1/4 mile), 400 hurdles, and 800. The 800's a bitch. You should hear my Sister give me hell about wearing tights or tripping over a hurdle.

Well, Eric class is over, time to go home + let Mya out to potty. Write back + stay cool. See you soon.

Your Home Boy,
Brandon McKenzie

Eric,
April 13, 2003

Thank you for writing me. It's a testament to my self-centeredness that I haven't written you before today. But two things happened which makes it easier for me to write you. First, your letter gave me an idea of what you were thinking (I didn't want to rain on your parade if you were all gung-ho about taking out Saddam.) Second, the war seems to be going well, all things considered, and many of my fears have been relieved. Still, I apologize for not writing you before now. At least Mom & Dad, Kim, and Grandma have been writing you regularly. You've got to be getting more letters than anybody else in the Marine Corps. Our family really loves and cares for you, myself included, although my behavior probably doesn't show it.

I'm sort of a pacifist, and a few weeks ago I was really concerned about this war thing. I just didn't see the compelling reasons to send our troops overseas to depose somebody else's dictator, especially when one of those troops was my brother. But I could see W's point, and there is at least a part of me that thinks that our government knows a little better what's going on in the world than I do. I'm glad it's not me making foreign policy decisions. But this sets a dangerous precedent, and I worry about us attacking Iran or North Korea, or whoever else we don't like. To me, the biggest improvement we could make would be to force Israelis and Palestinians to quit killing each other, and set up a 2-state deal. I think that's the main reason that the Arabs hate us (supporting Israel). That, and supporting the Saudi and Egyptian dictatorships, just so we have a stable source of oil. Anyway, enough foreign policy talk.

You said something in your letter about getting your life together. I know I shouldn't compare, but it looks to me like your life is already much more together than mine. I have so many things to be grateful for, but I'm still pretty unhappy sometimes. I've achieved some measure of material success, but "things" are so shallow and meaningless to me. I guess I always thought that if I got all the stuff I was supposed to get, then meaning and satisfaction would follow. If I had the house, the job, the car, etc., then I would be OK. But it doesn't work that way. It comes from the inside out.

Now for a news headline. I have always been jealous and afraid of you, because you have what I wanted more than anything else – the ability to meet girls, self-confidence, etc. It really surprised me to hear you say in December that you saw any desirable character traits in me. I recall the time you, Rick, and I went camping and riding in Uwharrie. I had a really good time, and I feel like you did, too. One of the reasons I used to hang out with Rick was that he liked me more than I liked myself (literally). The way that Rick portrayed me to you was much better than I could have done. I felt like Rick did such a good job of telling you about who I used to be, or who he thought I was, that I didn't want to hang out with you any more for fear that you would find out who I really was. And I knew that I was really much different than Rick portrayed me (boring, fearful, etc.).
So, in the past, when you would suggest that we go out sometime in Greensboro, I was terrified. I am afraid of bars. I am afraid of clubs. I am afraid of talking to girls. I am even afraid of talking to most guys. I thought it was better for you to continue to think of me as at least somewhat cool or normal, than for you to spend time with me and realize who I really was. Of course, the thought never occurred to me that you might interpret my refusals as meaning that I didn't want to spend time with you. That wasn't the case. I guess all this sounds pretty negative. But this is how I always thought (before counseling), and how I sometimes think now, when I get crazy. I wanted to tell you some of this in person, but the time never seemed to be right.

So, on to better things. My job is going very well. One of the guys quit, and we had to hire a replacement, who seems to be really good. I travel four days a week, and spend one day in the Asheboro office. It's like only having to work four days in a week.

Kitty is doing fine. My Greensboro house is fine; it needs about $4,000 in exterior repairs and painting, but I got a fat tax refund check, so I've got the money. I sold one of my three rental houses. I can't believe somebody actually bought it. I'll have to show it to you sometime. The other two houses are still rented out, though one tenant wants to leave because of the rats, and the other tenant lost most of he Housing Authority rent subsidy, so I'll be surprised if she can make her rent payment for long. But as o now, the rental house scene is calm. As soon as one of my two houses becomes empty, I'll probably pu it up for sale.

I have known since December that I need to change my living situation. It is no longer healthy for me t live by myself. It gives me too much time to be lonely. But I haven't quite figured out what to do. I may rent or sell my house, although I rather like my house, and don't want to get rid of it. But it's just too boring living by myself. The best thing I can come up with is, I have a friend in Winston Salem wh lives by himself, and I could move in with him. We get along pretty well. I don't particularly want to live in Winston-Salem, but I could handle it temporarily. It would be worth the move if it made me feel less isolated.

Well, I just got back from church. Dad goes every Sunday in Greensboro. I used to really like going to this church, but then I had a bit of a philosophical disagreement with some of the church's teachings. S I quit going, and went to a Sunday morning twelve-step meeting instead. I got a lot more out of the meeting, spiritually speaking. But I enjoy eating lunch with Dad afterwards, so sometimes I still go. Today the preacher was talking about how some people said that Jesus was the only way to get to heaven, but others said that Jesus was one of many paths to heaven (or enlightenment), others including Buddha, Krishna, Mohammed, etc. Of course, to be a Christian you have to believe that Jesus is the only way. At that point, I said to myself "well I guess that counts me out." My favorite bumper sticker is "God is too big for any one religion."

I bitch about a lot of stuff, but one thing that I am truly grateful for is my spiritual philosophy. It means a lot to me, to have a Higher Power which is completely non-judgmental (i.e., Hell does not exist), and to know that the only thing stopping me from a life of serenity and peace is my own character defects (fear, self-centeredness, resentment, pride, etc.) Sometimes, it's spooky to hear how similar the teachings of Christianity and 12-step programs are. I could talk about this shit forever. Maybe we can talk more in person about such things.

What's it like for you over in Iraq? The media coverage is non-stop. We see pictures of tanks driving through the desert, in the cities, sporadic sniper fire, celebrating Iraqis, looting, etc. For a while, I was almost obsessed with the war (Mom & Dad still are). But I had to stop spending so much time keeping up with it; it wasn't healthy for me. Even though I may not completely agree with the government's decisions, at least they seem to how to implement their plans. It's one thing to have a weak argument for invading a country, but as long as the invasion is relatively efficient, and has a positive end result, I guess I'm OK with it.

I'll bet you will have lots of stories to tell when you get back. See if you can get me a little trinket from over there; that would be cool. But the main thing I hope and pray for, is that you come back safe, unexposed to chemical weapons, and that much closer to finishing your time.

Well, that's enough for now. In the future, I'll try to write you more often than once every 3 months. Take care of yourself, and I will see you soon.

With much love and respect,

David

ACKNOWLEDGMENTS

Before I begin, I would like to let you know that it took me five years to begin writing this book after my first idea to do so. Once I began writing in early 2008, I was a self-employed real estate broker working 80 hours a week on average. In addition, I was blessed to have added my beautiful daughter, Savannah Jean, to my life. Having said that, my devotion to this book was limited to spare time, naptime and slow times in real estate. So although unintentional, there is a small chance that I may have left someone out that may have had an impact on my book or me as it relates to the story. So for everyone in my life who has given me support and the courage to accomplish and achieve, thank you.

I would like to begin by thanking the Courier-Tribune of Asheboro, NC for publishing my address and my dear friend Abby for giving me a journal. Not only did you ask me to write in it, but also you went above and beyond by giving detailed instructions on what to do with it. If it hadn't been for you, I wouldn't have a journal and all of this would have been forgotten—just as it had been before I reopened the journal. I wish you continued success in all of your endeavors.

I would like to thank all of you who supported me and those around me during our deployment. Thank you for taking the time to write someone you have never met or may barely know. At a time such as this, it means more than you can imagine to have the support of someone other than those who support me no matter what. Thank you:

Sarah and the Franklinville First Baptist Youth Group
Carol and Dennis "Dink" Routh
Grace Foster Moore
Lauren Kilby
Leslie Morgan
Danielle Ledesma
Sarah Blakely and the Oakwood Park Baptist Church
Laurie and Jeff Sypole
Jim and June Davis
Betty Martin
Ernest and Marie Pugh
Evelyn Yow
Rodena McCorquodale and the Willing Workers Sunday School
Class of Bethany United Methodist Church in Franklinville, NC

Mrs. Beane's Kindergarten Class at Farmer Elementary School
Jessica Evans, 2003 student at South Asheboro Middle School
Julio Olivares, 2003 student at South Asheboro Middle School
Brittany Hooker, 2003 student at South Asheboro Middle School
Daniel Ching, 2003 student at South Asheboro Middle School
Michael Rosas, 2003 student at South Asheboro Middle School
Lee AnneWoodall and her East Middle School Spanish students in Biscoe, NC
Savannah Morgan at the Green Street Baptist Church
Michael and Christy Tyson and Kids at the Green Street Baptist Church in High Point, NC
Charles and Hilda Neill at the Green Street Baptist Church
The Friends in Christ at the Green Street Baptist Church
and Abby's roommate Jen

Thank you to all my longtime and dear friends who have always supported me and have stayed in contact with me through the many ever changing seasons of my life. You are some of my best friends. Thank you:
Aaron Guido
Katie and Cindy Myers
Patrick McGrath
Morgan McGoldrick
Colleen Farrell
Brandon McKenzie
Neil Watson
Shelly and Garrett Lacoss

I have been blessed to have a beautiful, caring and loving family. Unless you've married into our family, you have been there for either my entire life or your entire life and it will always be like that. In addition, you were there to support me or my immediate family during this time and you may have also been there during the writing process. Whatever your role may have been as it relates to this book, thank you:
Kim and her beautiful children, Joshua, Julia and Lindsey
Grandma Jones
Sam and Sansia Coble
David Cox
Randy and Lorraine Newsome and Laura
Larry and Kay Jones
Russell and Delores Cox
Papaw and Grandma Cox – In loving memory of Mavine Cox

Although our paths had not crossed before the settings of this book, you have since come into my life and have helped me in one way or another during the writing and publishing process. We have grown from acquaintances to great friends and for that, I am grateful. Thanks:

Carrie Barham
Amanda Seymour
Arthur Secor
William Muench, IV
Mike Khaldun
Anna Ray
Alex Birbach
Jeffrey Spencer
Erica Patron
Melissa Ricks
Sanjay "Sean" Gautam and his wife Magali

Ryan Russo, we met in the gym of Courthouse Bay in Camp Lejeune, NC trying to "Get Big." Well we're still trying. Tell your mom, Harriett hello for me.

Russell Church, we met as enemies in CA but now we're the best of friends at home in the Carolinas. Russ, tell your mom Susan hello for me and give that beautiful little girl of yours a kiss for me.

Who would have ever thought that we would have gone through and experienced the things we have, together, and still be the best friends? Thank you for being my best friends through all of our ups and downs. Semper Fidelis!

Lastly I would like to thank my heroes, Jimmie and Cynthia Cox.

Dad, no matter where I was or what day it was during the war, the one thing I could count on was you writing me almost daily and always shooting straight with me. Thank you for your strength, courage, wisdom and love you gave to me throughout this book.

Mom, it took me having a daughter and struggling with the terror of losing her to realize why it was so hard for you to write. Now that I am home safe you have gone to great lengths to help me through the writing process as my editor. Thank you for the care packages, support, ideas and love you gave me throughout this book.

I can't even begin to imagine what I have put both of you through. To me, you are the best parents anyone could ever ask for. All I can do is continue to show you how grateful I am, how much you mean to me and how much I love you. My greatest achievement would be to instill in Savannah, what you have in me. I love you!

ABOUT THE AUTHOR

Photo by: Alexandria Bright

ERIC J COX

Born in Greensboro, NC, raised in Asheboro, NC, Eric made Charlotte, NC, his home in 2004 where he began working as a bartender. He purchased his first home in 2005 and began purchasing other properties as investments. In 2006 he achieved his NC real estate license and became a Realtor®. Eric excelled as a commercial and residential investment specialist and was nominated for the Rookie of the Year Award as a multi-million dollar top producer following his inaugural year. Today he is the Broker – Owner of The CHARLOTTE Real Estate Firm

Most recently, he has completed this book as his first and begun work on his next book. He has opened his own micro-press publishing company, The Charlotte Press, where he hopes to help other writers and authors publish and market quality books. Eric is also founding a non-profit organization to benefit veterans. A portion of the proceeds from this book's sale, among other donations, will go to benefit his charity.

Eric is a single father residing in Charlotte with his beautiful daughter Savannah, his dog Diezel, and his two cats Heidi and Chloe.

Your feedback and/ or comments are greatly appreciated at
Cox3960@Gmail.com or by mailing The Charlotte Press;
229 S Brevard St, Suite 300; Charlotte, NC 28202

www.ingramcontent.com/pod-product-compliance
Lightning Source LLC
Chambersburg PA
CBHW031245090426
42742CB00007B/322